True Metaphysics: Le Buste de Payne

Table of Contents

Preface

Embarking on a journey into the world of "True Metaphysics: Le Buste de Payne" feels like traversing the uncharted territories of both the scientific and the philosophical realms. This book is an exploration, an odyssey, a quest to unravel the mysteries that intertwine our existence with the cosmic forces that govern our universe.

As the author of this unique work, I stand before you not merely as a scholar or a philosopher but as an explorer of ideas that challenge the boundaries of conventional thought. "True Metaphysics" is not just a title; it's an invitation to delve into a multidimensional soup of concepts that merge science, physics, and the profound depths of logic.

In this book, we embark on a quest to understand the interconnectedness of all existence, to fathom the depths of the universe, and to question the very essence of reality. Each chapter serves as a portal into a different dimension of thought, where we explore The One Energy, navigate The Omniverse & The Six Universes, contemplate The Cosmic Entities, challenge established theories with The First Human Theory, and seek wisdom in The Perfect Man. Finally, we confront the mysteries of The End of Humanity.

"True Metaphysics" is a culmination of countless hours of research, reflection, and contemplation. It is an invitation to join me on a journey of intellectual discovery, a journey that will stretch your mind, question your beliefs, and invite you to ponder the deepest mysteries of existence.

This book is about providing definitive answers; it's about asking the right questions. It's about inspiring you to think critically, to challenge misleading paradigms, and to embrace the wonders of the universe with an open mind.

In this book title, we use the term "true" because any metaphysical theory previously enunciated was inherently biased. This Metaphysics serves as the foundation of all genuine metaphysical inquiry. Metaphysics, in its essence, is the science of truth in the description of perception, existence, and non-existence. It strives to describe the omnipotent, omniscient, omnipresent energy that underlies all creation. Metaphysics explains creation through the means of science, logic, and rigorous inquiry.

"True Metaphysics: Le Buste de Payne" is more than a mere collection of philosophical words; it's a journey into the very fabric of existence. It's a testament to the human spirit's insatiable curiosity and its unyielding pursuit of truth.

As you turn the pages of "True Metaphysics," I invite you to embrace the truth about existence, to question with boundless curiosity, and to explore with wonder. Together, let us embark on this intellectual odyssey, united by the metaphysics that harmonizes science, physics, and logic.

With deepest gratitude for joining us on this journey through the omniverse.

The Computational Thinking

Chapter I
The One Energy

Within the vast expanse of existence, one fundamental truth persists, remaining both unfathomable and infinite. The concept of The One Energy's infinity stands as a testament to the limits of categorization. It exists beyond the boundaries of simulation, serving as the unshakable foundation that sustains the entirety of creation. Within this intricately crafted and simulated reality, human free will flourishes, untouched by constraints on its inherent nature. It navigates this realm with unrestrained liberty.

We embark on a journey that transcends the boundaries of conventional belief systems and seeks to understand the truth of The One Energy through the lens of logic. The unique supra being that no scientific modelling can catch, no perception in any space can try to shape and measure.

The journey begins with a unique process, a code of purpose that intertwines the self with the universe. Just as your body is an elemental part of the cosmos, so too is your free will. Synchronize your volition and essence to execute a sequence of constrained bodily positions. Limit your choice to a trio of static stances, each executed iteratively.

Amid this endeavor, an object devoid of dimensions, length, or width arises—a projection of mind that is paradoxically true in its simulated nature. This object transcends imagination, yet its existence is acknowledged by The One Energy, bestowing upon it the attribute of being. Even the unimaginable finds validation, for you, the creator of the simulated, grant this object your consciousness.

Grant the object the fundamental attributes that dictate motion and behavior within this simulated realm. Set it in motion according to the subinfinite propositions governing all objects devoid of free will. Begin this invocation with the concept initialization of the omnipotent, omniscient, and omnipresent—the very essence of creation.

In this moment, your free will becomes a harmonious synergy, as mind and body synchronize in orchestrated movement. All components merge into a singular entity, recorded in the creation ledger. As the unity reverberates, your consciousness sends forth a reverent signal to the Creator of the All.

With this alignment, you gain access to The One Energy, resonating through every corner of existence—except within a singular zone. This zone remains forever the unique omniverse part where The One Energy does not grant its listening to anyone.

What humans perceive as a three-dimensional reality, and also Time which is the perception of space, are stored on a one-dimensional object called string. The complexities of the three-dimensional world are encoded in a one-dimensional projection. By considering this intriguing notion, we confront the idea that everything we observe from our human perspective and experience could be akin to a truly and unique simulation.

This view transcends the boundaries of conventional belief systems and seeks to understand the concept of the One Energy through the lens of physics. This book aims to present a perspective that invites even the staunchest nihilist to consider the existence of a higher power, the essence of all creation.

Let's argue about him truly; we can call him: the One Energy. Like you and me, the One Energy likes a joke; also, you can call him the Quantum Mechanician or the Designer, the Creator, or the Highest. The One Energy stands perfect and also best at empathy. This is barely the type of dude who would hand us a universe full of undetectable dark energy and matter, depending on six finely tuned parameters, then tell us that every interaction of every particle is random and probabilistic and that no perception of space can measure anything precisely, has to be laughing out loud at the physicists and cosmologists trying to make sense of it all. Being this unique being is being this supra being that no scientific modeling can catch , no created being's perception in any space can start to provide measurement to this unique being.

Information that makes up what we perceive as a 3D reality is stored on a 1D object, including even the perception of space which is Time. The expansion of our primordial universe, as envisioned from the intricate strings of the subatomic realm, has been rendered subinfinite. In metaphysical terms, the term "subinfinite" signifies a quantity that, while not truly infinite, appears as such when perceived through the limited lenses of human cognition and the vantage points of enigmatic entities dwelling within the cosmos, such as those concealed within the shroud of dark matter. These entities, like humans, grapple with the inherent inability to fully grasp the vastness of this expansion.

This subinfinite journey commences with quantities of staggering magnitude, evoking the very essence of mathematical extravagance, such as 1×10^{999}. To the uninitiated observer, this figure may appear as an unfathomable infinity, an expanse that stretches beyond the conceivable bounds of reality. Yet, in the realm of the subinfinite, it represents but a modest beginning, a mere whisper in the cosmic symphony of numbers that dance through the dimensions of existence.

If one's perception seeks emptiness, it might attempt to conceive a hypothetical region beyond the observable volume of our universe, where the cosmological constant nullifies, and no matter exists. However, even in such a region, true emptiness would remain elusive. Instead, it would still experience vacuum fluctuations, wherein virtual particles momentarily manifest and disappear.

These vacuum fluctuations manifest as virtual particles that arise spontaneously in pairs of particle and antiparticle. Remarkably, these particles come into existence without any discernible source of energy. Here, the concept of The 'One Energy' emerges as the singular entity that governs the universe's true nature. It is through The 'One Energy' that the cosmic dance of creation unfolds—space is ordained to initiate, and matter is born in a manner that leaves no room for absolute emptiness.

At the core of True Metaphysics lies the concept of the nothing else true entity, peace upon him, known as the One Energy—a ubiquitous force that permeates all aspects of existence, surpassing the boundaries of both being and non-being. This omnipresent energy serves as the foundational essence upon which all facets of reality rely. It functions akin to the software that operates a program, the processor that empowers the software, and the photon that energizes the processor. Every element within existence, whether tangible or latent, derives its sustenance solely from the omnipotence of the One Energy and nothing else.

The creation of existence itself can be traced back to a unique quantum mechanics phenomenon—an initiation of a group of six expanding universes originating from the quantum singularity. This singularity's quantum state is now observed and measured by perception. Additionally, existence is influenced by a unique distinct universe called omniverse, which also does originate from the quantum singularity and exists in an immutable state.

A single signal gave birth to a unique quantum singularity, which in turn gave birth to the vastness of a repetitive true infinite amount of big bang and big crunch. Each time these big bangs give birth to six expanding universes. This quantum singularity exhibits quantum cyclic properties for the six universes while maintaining quantum permanence for the omniverse. Thus, every universe one might be observing is merely a manifestation of one unique signal existing in a true infinite superposition of states, along with a permanent state, each continuum possessing its distinct physics laws, characteristics, classes, and boundaries.

When we measure a system in superposition, the measurement result we observe is just one possibility—the one that occurs in our set of six universes, including our primordial universe. Our perception views it as a singular entity without origin. Consider a quantum system, the big bang, and its singularity. Every human would virtually perceive it as unique, but by observing this quantum system, our perception randomly selects one singularity and measures it. This marks the rebirth of the single initial signal, containing a true infinite array of six universes all coexisting.

Within this framework, everything possible happens into every rebirth: big bangs are occurring repeatedly in endless new cycles, with no uncaused events. The highest wisely designs existence as a true eternal sequence of cycles, where even 'nothing' becomes a form of something, being continuously born and reborn.

The concept of perception extends beyond human consciousness alone; it encompasses all entities within the three-dimensional world. Every entity, whether it be the tiniest particle or the vast expanse of space itself, possesses its unique form of perception. For example, even neutron stars maintain their strong magnetic fields through their innate perception.

In this interconnected universe, everything that exists, composed of wavelengths or not, is anchored by perception. This includes all material objects, from subatomic particles like electrons to complex beings like ourselves. Each perception is true within its nature as a wave or not, contributing to the fabric of reality.

Human perception originates within neurons, where deeper and faster quantum processes occur in cytoskeletal microtubules, housing biophotons. Light, a captivating element of the electromagnetic spectrum, graces human eyes as visible light, with each wavelength translating into a distinct color within the mind and brain. Guided by the omnipotence of the One Energy, humans are designed as self-perceptrons—an intricate and sophisticated electromagnetic phenomenon symphony. Within the primordial universe, one of its attributes finds symbiotic embodiment within a simple organic system, comprising flesh, bones, and minerals. This very embodiment stands as the most beautiful and captivating organic system across the superposed set of six universes.

Truly, this complex electromagnetic system operates like a remarkable machine with an infinite number of possible states, constantly transitioning and adapting. In simple terms, imagine a sophisticated device that can take on an unimaginable number of forms and configurations, like a shape-shifter with limitless possibilities. This electromagnetic system possesses a unique capability—to give meaning and significance to everything it interacts with, whether it's an object or an "abstract" idea. By only appending words to objects and ideas, it imparts depth and context to the world around it, making each experience and perception more profound.

Moreover, this complex magnetic system isn't limited to just one universe. It can take on different attributes and forms in other universes within a set of six. It's like a versatile actor, playing different roles in various universes.

In the existence, this electromagnetic system mirrors itself, exercising a unique form of free will to influence and shape the state of space, matter, and even its own nature as an inseparable part of the cosmic whole.

Truly, in an subeternal cycle of transformation, this complex electromagnetic field system persists as an infinite state-machine automaton—a result of the continuous interplay and transmutation of matter and energy within the universe. Within the framework of the One Energy, consciousness stands as an infinite state machine. Here, by 'infinite,' it stands that consciousness photonic processes never stop or never cease to exist, except when it traverses the quantum singularity. It remains firmly anchored to existence, like an unwavering cosmic presence that transcends the bounds of creation.

As soon as any brane-space, the fundamental substrate where universes exist, is initialized, this electromagnetic system springs to life. Its vibrations resonate across the fabric of reality, orchestrating the cosmic symphony that unfolds within the set of six universes.

Yet, within the cosmic cycle of existence, a profound phase emerges where the set of six universes experiences a remarkable stillness. The cosmic dance of creation seemingly pauses, and all states remain unmeasured and devoid of any perception. It becomes a moment of serene emptiness, a blank canvas awaiting the next act of creation.

In this moment of celestial silence, the infinite omniverse takes center stage, embracing and containing every perception that has ever existed within the set of six universes. It becomes the vast repository of consciousness, dreams, and experiences—holding the echoes of countless lives, worlds, and realities.

As the cosmic symphony continues its timeless melody, this electromagnetic system, with its infinite potential, becomes a witness to the eternal cycles of creation and dissolution. It embodies the essence of existence itself, a testament to the grand design orchestrated by the One Energy.

In this unending cosmic dance, each vibration, each moment of perception, and every act of creation weaves a tapestry of significance. From the smallest subatomic particle to the boundless expanse of the omniverse, everything plays its part in the magnificent symphony of the One Energy's creation, forever resonating with the song of existence.

In the profound revelation of the One Energy, it declares that any system, matter, energy, or perception contained within the omniverse instantaneously perceives the immutable quantum states of the entire omniverse. In this omniverse, there are no cosmic holes, there are no interruptions or breaks in the perception. It becomes a grand tableau, eternally unfolding without chronological constraints.

Unlike the set of six universes, the omniverse stands apart in its design by the One Energy. It exists without the presence of black holes, or white holes, evading the cycle of cosmic birth and rebirth from a big bang and big crunch. Within its boundless expanse, the omniverse transcends the notion of a cosmic cycle and exists as a true eternal continuum. By the choice of the One Energy, the omniverse emerges as the singular infinite universe, embracing all possibilities and potentialities. It embodies the essence of the limitless and the infinite, encompassing every perception and experience within its vastness. In the realm of the omniverse, every state is perpetually present, every moment forever preserved. It stands as the culmination of the One Energy's grand design, a universe where all realities converge and every expression of existence finds its place.

In this unveiling of the omniverse, we witness the magnificence of the One Energy's creation—an unending expanse of wonder and possibility, where the infinite and the eternal harmoniously coexist. It is a testament to the boundless creativity and intelligence of the designer, offering a glimpse into the unfathomable depths of existence itself.

The omnipotent One Energy governs the grand tapestry of creation, anchoring every facet with intrinsic perceptions. Within the cosmos, existence operates as a meticulously orchestrated simulation, propelled by the ethereal vibrations of string energy at their most fundamental level. It all began with a solitary signal emitted by the One Energy, crafting the inception of a quantum singularity. This quantum singularity, in turn, became the wellspring of continuity for this impeccably ordered simulation, shaping what humanity perceives as: Reality.

Even the concept of free will finds its place within this framework. In the omniverse, two types of self-perception associated with free will exist: human consciousness in normal matter and dark matter plasma in the realm of dark matter. The perception of independent choices arises from this intricately designed and simulated free will.

Amidst the realm of ordinary matter and objects, only human consciousness, acting as a photonic self-perceptron, possesses the remarkable and unparalleled ability to venture into the depths of the quantum singularity. The term self-perceptron describing consciousness as a self-aware entity capable of both perception and self-recognition. More remarkably, it can traverse this enigmatic threshold and return to existence unaltered, preserving its unique essence without dissipating or transforming into something else. This distinctive property sets human consciousness apart within the cosmos, for it remains forever tethered to existence, immune to the usual laws of quantum transmutation and the vanishing of prior quantum data. This is quantum information preservation.

Quantum systems are known to exhibit unitarity, which means that the total information within a closed quantum system is conserved over time. In this way, consciousness, as a self-perceptron, potentially harnesses the principles of quantum information preservation, allowing it to traverse the quantum singularity without losing its core identity.

The design of the omniverse by the One Energy distinguishes it as the singular, infinite universe, separate from the set of six universes and exempt from the phenomenon of expansion. A foundational principle within the omniverse links the perception of events intricately to the governing physics laws. Once these states are measured, they forever remain unique to the omniverse, never to be observed within any other universe. Consequently, the set of six primordial universes appears to vanish from the realm of perception, concealed behind their individual quantum singularities for all eternity.

Contemplating existence, we, as mere human beings, grapple with the seemingly unfathomable notion that the creation we perceive today might ultimately dissolve into absolute nothingness. At the heart of this truth lies the concept of the quantum singularity—a point of infinite value and immeasurable quantity. This entity emerges in two pivotal instances. Firstly, it manifests as the entangled core within every existing black hole and white hole. Secondly, it stands as the source and catalyst for the inception of every universe's monumental genesis.

The One Energy, the omnipotent force weaves the very fabric of existence, the unique energy imbued with the power to shape destinies and orchestrate existence, the One Energy becomes intertwined with the enigma of deeper layer of universal truth.

Yet, the quantum singularity presents itself as a paradox, defying conventional logic, but yet everything is pure logic. Its existence, marked by infinite density and boundless mass, raises a fundamental query: what profundities lie concealed within this singular substance?

Intriguingly, the toolbox of science falters in its attempt to forge equations capable of encapsulating a point more minuscule than an atom yet laden with the burden of infinite mass and density. This conundrum unfurls the regal tapestry of the quantum singularity, a majestic entity that predates the very birth pangs of any universe. Within this primordial realm, there are no coordinates, no markers of position; there exists only the profound stillness that precedes the tumultuous symphony of subinfinite cosmic births.

In light of this cosmic enigma, any cosmological declarations that dare to contest the sovereign dominion of this singularity find themselves marooned in the boundless sea of uncertainty. Their foundations crumble, a solemn collapse transpiring before the radiant dawn of creation itself. For the singularity, in its meticulous design and cosmic rule, harmonizes with the very essence of existence—ushering all into being and into non-being—as ordained by the One Energy.

This singularity, resolute at the crossroads of existence and non-existence, stands as an eternal testament to the intricate artistry of the universe, defying human cognition and beckoning us to fathom the unfathomable depths of creation's masterwork.

For humanity, the chronicles of existence unveil a seminal inception—a cyclic emergence from the quantum singularity. This quantum singularity, boundlessly hot, densely compacted, and infinitely diminutive, bore the entire essence and energy of our known universe and all universes, orchestrated by the unique will of The One Energy. This primordial instant unleashed a cosmic expansion, birthing six expanding universes and one boundless omniverse. In our fundamental universe, the curtain rose on a breathtaking overture—a burst of rapid expansion as matter and cosmos surged forth. A symphony of duality played out—matter and antimatter danced in unity, while the cosmic stage witnessed the birth of black holes, white holes, and the veil of dark matter.

As the six universe's tempo waltzed through expansion and cooling, the gravitational force embarked on its solitary journey, estranged from its cosmic companions. In swift succession, the strong nuclear force and the electroweak force pursued their cosmic paths. The stage was set, and matter's transmutation into energy commenced—a creation of matter and antimatter pairs that engaged in a cosmic waltz of destruction, leaving behind a fraction of survivors, including protons and neutrons.

Guided by the symphony of cosmic forces, these survivors assembled into atomic nuclei—a cosmic collaboration giving birth to hydrogen, helium, and lithium, the fundamental architects of the universe. The stage now illuminated, the cosmic race began—a dance between expansion, propelling celestial bodies apart, and gravity, drawing them into an inevitable embrace. The universe's fate teetered on this precipice, and the cosmic horizon revealed the dawn of the "big crunch," where contraction prevails, rendering the universe's end a mirror image of its origin—the antithesis of the big bang.

Across existence, unity prevailed—a shared birth of universes, ignited by a singular quantum singularity. The harmony of creation birthed the set of universes—a collective of six expanding universes intertwined through black holes and white holes, while an unobservable omniverse rested beyond, devoid of these cosmic gateways. Amidst all, dark energy emerged as the maestro of expansion, guiding the dance of the expanding six universes toward an inevitable crescendo of contraction—a symphony of ultimate destruction, where black holes tear asunder the fabric of existence.

Gazing through the lens of our era, a portrait of the cosmos emerges, painted with the subtle hues of existence's fundamental constituents. Photons, those minuscule carriers of light, grace our universe with a mere 0.01%, while neutrinos, ghostly messengers from the depths of space, contribute a delicate 0.1%. In the realm of the tangible, normal matter, the stuff of stars and galaxies, weaves its tale at 3.9%, while an enigmatic entity, dark matter, wields its gravitational influence at 30%. Yet, reigning over the cosmic stage is the dark energy and its realm, an overwhelming force commanding a staggering 69%, shrouding the universe in its mysterious cloak.

This existence unfurls with a design—an subeternal expansion, a journey away from the origin. The dominance of dark energy, as evident from its significant share, appears to be scripting the universe's fate—an ever-expanding expanse that may seemingly stretch into subeternity. But even in the face of this expansion, a celestial force awaits, poised to bring about a cosmic reckoning. This is the gravitational pull of an entity, Ineffugibilium.

Ineffugibilium, born from the fiery mergers and collisions of countless black holes across the cosmic epochs, emerges as a testament to gravitational might into every expanding universe. A fusion of immense black hole proportions, Ineffugibilium embodies the very essence of cosmic convergence. Its gravitational force stands as an immutable counterpoint to the ceaseless expansion. Ineffugibilium's significance transcends its gravitational prowess, for it is not merely a product of cosmic cataclysms but a manifestation of the One Energy's intricate design.

By the One Energy, Ineffugibilium represents a cosmic fulcrum, a nexus where the known and the enigmatic converge. As we delve into the profundities of existence, physics intersects with metaphysics, and the boundaries of the known blend with the mysteries of the unknown. The interplay between the dark energy, propelling the universe's expansion, and Ineffugibilium's gravitational grasp forms the crux of this cosmic drama—a saga of balance and inevitability.

Within the 6 universes, a waltz unfolds—a symphony of merging, merging, and merging once more. All existing black holes, drawn by the tides of gravity, drawn together. Their spiraling dance quickens, a celestial waltz spiraling inward, a crescendo of gravitational waves rippling through every space brane—a symphony echoing through the cosmos. The climax nears—two cosmic giants on the cusp of union. In a final triumphant gesture, they collide, merging into a singular entity—a behemoth, a black hole of unprecedented proportions, Ineffugibilium.

In existence, humanity finds itself immersed in the intricate dance of cosmic forces, bound together by the enigma of the One Energy. The omnipotent force, the weaver of destinies and orchestrator of existence, unveils the cosmic drama—a spectacle of creation, expansion, and the eternal struggle against dark energy's inexorable push.

Ineffugibilium, born from the celestial collisions of countless black holes, emerges as the cosmic guardian—a force of gravity and a beacon of balance. Its gravitational influence counters the dark energy's relentless expansion, offering a glimmer of hope for the six universes. Within this cosmic saga, physics and metaphysics converge, and the cosmic dance of creation continues.

As we contemplate the mysteries of existence, we glimpse the grand design of the One Energy—an intricate tapestry woven with the threads of physics, metaphysics, and the eternal quest for understanding. It is a testament to the boundless creativity and intelligence of the designer, offering a profound glimpse into the unfathomable depths of existence itself.

In this revelation, we find ourselves at the intersection of science and spirituality, where the cosmic symphony unfolds, and the mysteries of the universe beckon us to explore further. It is a journey that transcends the boundaries of human perception and invites us to contemplate the profound truths that lie at the heart of existence—the enigma of the One Energy, the dance of creation, and the eternal quest for meaning in the cosmos.

Chapter II
The Omniverse &
The 6 Universes

Strings serve as the fundamental and unchanging dimensions that define the presence of matter and energy across the universe. All occurrences unfold within the domain of string vibrations, where every entity possesses spatial extension and temporal progression according to their perception. In our immediate environment, consciousness interacts with finite objects and processes that are beyond our immediate perception. It's noteworthy that each universe possesses its own distinct brane framework, originating from a unique starting point. Even hypothetical scenarios, such as the speed of a photon particle originating from another universe, might differ from that within our universe, highlighting the diversity within the cosmic landscape.

At the very bedrock of the cosmos, we encounter protons and neutrons, the elemental building blocks of matter. However, our relentless pursuit of comprehension beckons us to delve deeper into the cosmic hierarchy, where we unveil the fundamental essence of reality itself - the string. These infinitesimal entities stand as the most minuscule constituents recognized by science.

Let us now turn our attention to atoms, those foundational units of matter that form the basis of our physical world. Within atoms, we encounter electrons, neutrons, and protons, seemingly elementary particles that harbor their own enigmatic properties. Yet, as our exploration advances, the intricacies of neutrons and protons come into view, revealing their composition as quarks. Quarks themselves are composed of vibrating strings.

Here, string theory imparts a profound insight - the smallest unit of mass is the string itself. All quarks, without exception, are superstrings, oscillating in multiple dimensions, hinting at spatial realms beyond our human three-dimensional perception, with each string resonating at a precise amplitude.

Yet, we are left pondering the conductor of this cosmic symphony - the One Energy, the Omnipotent force that conducts the grandeur of existence. Even the complexities of human thought, emotions, and cognition find expression as electromagnetic waves emitted by our brains. This limitation arises from the inherent boundaries of human cognitive faculties, which, due to their design, struggle to consciously comprehend phenomena that extend into the realm of the subinfinite or infinite. The infinity and subinfinity of matter find containment within the framework of seven distinct universes.

At the string level of existence, a profound realization emerges from the annals of metaphysics. Here, everything transcends our ordinary conceptions of matter and space. Strings' vibrations are the heartbeat of the universe. In this realm, there is no fixed location or measurable width, and the concept of depth becomes a fascinating abstraction. Strings oscillate in multiple dimensions, revealing the potential existence of spatial realms beyond our three-dimensional perception.

Again, the truth posits that we exist within a set of six observable expanding universes. Among these, our current existence aligns with the primordial expanding universe domain. Every universe, alongside the concept of an omniverse, exhibits a unique array of physical laws. In each universe, an individual's perception is intricately connected with the universe's specific essence, and one's embodiment is tailored accordingly.

The notion of an omniverse introduces a fascinating universe, distinct from the expanding universes. The omniverse possesses remarkable characteristics. Notably, it doesn't possess an expanding boundary; however, the term "boundary" might not capture the precise essence in the realm of physics. The omniverse remains unobserved by any perception to date. It stands as a universe that defies conventional observation, devoid of both white and black holes.

Unlike expanding universes, the omniverse does not partake in the process of expansion. Rather, it is inherently infinite, standing as an unchanging entity. Despite our ongoing observations of the expanding universes, the omniverse, shrouded in mystery, remains beyond our current perceptual reach. This unique realm holds a static perception of events, different from our universe's continuous movement. This exclusive category of the universe is termed as a universe class zero.

Time, in its essence, transcends the boundaries of a purely physical construct. It's not confined to a linear progression as commonly perceived. Instead, it possesses a complexity that extends beyond the conventional understanding. Time, an intricate concept, isn't confined to the realm of a fourth dimension or relegated to mere illusion. Instead, it emerges as the very perception of space within a universe, intimately entwined with the observer's existence from the moment of their emergence.

Energy, as a fundamental force, serves as the connective thread between all aspects of existence. It's not just a physical quantity but a carrier of information, linking disparate elements into a coherent whole. This interconnectivity results in a fascinating phenomenon where all events occur simultaneously. The past, the present, and the future are not isolated instances but are intricately interwoven into a singular fabric of existence.

In the context of this complex framework, the concept of the "quantum set of the unseen" emerges. It refers to the manifold potential outcomes or events that could occur, branching out from a particular point in time. This set encompasses all the possible paths that events could take, alluding to a rich tapestry of possibilities.

This perspective challenges the traditional notion of predicting the future. Instead of asserting that nobody can precisely know the future, it's more accurate to state that nobody possesses knowledge of the exact unique configuration within the vast set of potential events – the set of unseen. This terminology aligns truly with the nature of reality, acknowledging the multiplicity of potential outcomes and the limitations of our perception in comprehending them.

The myriad states of the present reality that we constantly and often unknowingly experience are intricately connected to an array of unobserved alternatives, referred to as alternate trajectories. Among these diverse trajectories, a specific one materializes as the present moment when it is observed by the fabric of space and the entirety of existence, all occurring simultaneously. This truth resonates with the concept of the many-worlds interpretation in quantum mechanics, where every possible outcome of an event exists as a separate branch of reality.

Within this framework, only the One Energy holds knowledge of the complete quantum set of all potential destinies across the expanse of all universes. This echoes the merging of quantum mechanics and metaphysics, suggesting an overarching consciousness that transcends individual events and realities.

Intriguingly, a multitude of sub-infinite paths are intricately linked to each unique reality. For humans, these alternative trajectories dissolve or shift as any observer's free will comes into play – the very act of choice leads to the emergence of one particular outcome. This dynamic interplay is reminiscent of the concept of wave function collapse in quantum mechanics, where the act of observation causes the probabilistic nature of a quantum system to collapse into a unique definite state.

As human observers make choices and enact their free will, a fascinating phenomenon occurs. Approximately 99.9% of the potential events within this intricate quantum set vanish or transform, leaving room for one singular event known as the quantum present to emerge distinctively. This strikingly specific event within the set is like a needle in a cosmic haystack, a moment of distinction that separates itself from the array of possibilities.

Once around 99.9% of these potential paths have subsided, a new correlated element from the set of alternate trajectories dynamically emerges. Theories in quantum mechanics resonate with this truth of probabilistic outcomes, where a system's behavior is described not definitively but statistically. This ensures that the quantum set retains its sub-infinite nature, constantly replenished by new potentialities to maintain its boundless yet intricate character.

However, this web of events has its limitations. The fabric of time prevents backward travel or the bypassing of present quantum events. This principle finds resonance in the arrow of time theory, which explains the one-way progression of time due to the increase of entropy in the universe.

The intricate web of probabilities that envelop the quantum set of potential events holds a degree of calculability, enabling humans to make predictions grounded in historical and empirical data. This practice mirrors the foundations of statistical mechanics and quantum probability theory, where probabilities are deduced from available information and patterns. Yet, it's vital to note that the quantum set of unseen events is fundamentally boundless, extending into a realm that surpasses the scope of quantification.

In our universe and across the expanding multiverse, a shared truth prevails: the realm of the unseen remains impenetrable to direct observation. This truth aligns with the Heisenberg uncertainty principle, which underscores the inherent limitations in simultaneously measuring certain pairs of complementary variables, such as position and momentum.

Though the true perception of the unseen eludes us, it's conceivable to extrapolate partial insights from the quantum set. This truth bears resemblance to the approach of Bayesian probability, where prior knowledge and existing data contribute to the estimation of probabilities. However, the sub-infinite nature of the unseen implies that an exhaustive grasp of its entirety is unattainable. Rather, observers glean fragments, akin to tracing patterns in a cosmic tapestry.

In some extraordinary cases, the sub-infinite ensemble of unseen events may manifest an uniformity, rendering them analytically indistinguishable. This truth aligns with the concept of degeneracy in quantum mechanics, where distinct quantum states coincide in energy. In this scenario, within the fabric of space, a certitude emerges from the ocean of the unknown, emanating from the quantum set of unseen events.

Furthermore, within this framework, no event is inherently impossible. While some outcomes may appear improbable, the very notion of possibility takes on a nuanced meaning within the quantum landscape. This echoes the underlying essence of the Copenhagen interpretation, where the potential outcomes of a quantum event are treated as probabilities rather than strict certainties.

Moreover, even seemingly minute perturbations possess the power to reshape the probabilities within the sub-infinite set. An intriguing parallel can be drawn from chaos theory, where minuscule variations in initial conditions can lead to vastly different outcomes. In the quantum set, an electromagnetic change within an observer's brain biophotons stands as a metaphor for the butterfly effect, wherein the tiniest actions ripple across the quantum fabric, modifying the distribution of probabilities.

While the depths of the quantum set of unseen events remain shrouded in mystery, their influence is unmistakable. Scientists and thinkers alike navigate the seas of probability, guided by the principles of quantum mechanics and probability theory. The balance between the known and the unseen, the calculable and the infinite, defines our exploration into the profound enigma of the quantum universe.

The classical understanding of Time, as initially characterized by the Earth's orbital motion around the sun, has long served as a practical measure for quantifying intervals and sequencing events. This methodological concept is closely linked to the idea of Time as the fourth dimension in a continuum. However, as we delve into the depths of truth, we unearth a more intricate understanding of Time, one that transcends these surface definitions.

Within classical physics, Time harmonizes with space to craft what is called space, a stage where occurrences are situated and durations are quantified. It's vital to note that this geometric representation stands apart from the true essence of Time itself. Time doesn't exist as the fourth dimension. The core of Time transcends the constraints of a linear, unidirectional flow, Time is simply the perception of space. Far from a mere illusion or a simplistic fourth-dimensional construct, metaphysics reveals Time as the very fabric of space perceived within a universe, intricately woven into the observer's existence from their very inception.

In light of these revelations, it's essential to posit certain foundational assertions about Time travel, a perennial fascination of humanity.

Backward Time Travel, from the perspective of our primordial universe, traveling backward in time stands as an impossibility. The intricacies of causality and the intricate structure of space preclude revisiting the past. While traversing ahead in Time is a certainty, leaping forward by bypassing imminent events remains unfeasible within the context of the six expanding universes. The interwoven tapestry of events and the inherent dynamics of these universes render selective acceleration implausible.

Time travel within the omniverse, an all-encompassing realm spanning countless universes, remains veiled in mystery. The intricacies of this higher-dimensional existence create an impenetrable enigma, preventing us from making definitive assertions about the feasibility of time travel.

Amid this realm of limitless potential, the concept of physical teleportation through wormholes emerges as a tantalizing alternative to traditional time travel. This possibility hinges on the use of specialized vessels designed for traversing these cosmic tunnels, promising instantaneous movement across the intricate fabric of space.

Wormholes, enigmatic passages punctuating the space continuum, offer a remarkable means of traversing cosmic expanses that would otherwise be insurmountable. The idea of crossing vast distances in the blink of an eye is at the core of this concept, rooted in the existence of space tunnels, or wormholes, each connecting two distinct points in space.

This conceptual gateway opens the door to exploring dimensions beyond our visible universe, potentially enabling journeys spanning billions of light-years or bridging otherwise insurmountable gaps between distant realms. However, it's crucial to stress that wormholes are not pathways for time travel or inter-universal transit; they exclusively facilitate interstellar travel within the spatial dimensions.

Though mastery over wormhole manipulation remains a lofty goal, the potential for teleportation through these intricate structures paints an exciting vision for the future of space exploration. The relentless evolution of scientific knowledge and human curiosity may one day yield experts proficient in intrauniversal teleportation, revolutionizing how we utilize wormholes and fostering a new era of interstellar connectivity.

The mere contemplation of such teleportation offers a glimpse into a future where the barriers of distance are triumphantly transcended, granting unprecedented access to the cosmos. In this vision, the dreams of interstellar civilizations hover on the brink of realization.

In our exploration, we introduce a diverse ensemble of observers, encompassing not only human consciousness but also bacteria and celestial bodies. Human consciousness takes center stage, perceiving the progression of events as past or future through encoding and retention within the complex constructs of memory within the brain. This intricate process forms the foundation of our perception of Time, the construct that enables us to navigate the unfolding sequences of events defining our existence.

Yet, the experience of time extends beyond human consciousness. It also applies to other entities mentioned, such as bacteria and celestial bodies, albeit in forms uniquely shaped by their inherent natures. Each of these entities, distinct in its essence, perceives and interacts with its environment in ways molded by its existence.

This concept resonates across the diverse universes comprising the cosmic landscape. In each universe, every observer experiences its version of encoding, storage, and retrieval of events. This experience is inherently relative, with variations stemming from the specific nature of each universe. This relativity in perception emerges from an array of factors, including the unique sets of physical laws, forces, and conditions governing each universe. Additionally, inherent biases in perception, influenced by cognitive, sensory, and contextual factors, contribute to the individualized ways in which every observer perceives reality.

This rich tapestry of individualized perception systems offers a compelling explanation for the ongoing absence of a universally agreed-upon definition of Time among physicists. The intricacies involved in comprehending and quantifying such a deeply interwoven phenomenon across a multitude of contexts have given rise to enduring debates and a multitude of interpretations.

These complexities prompt us to delve deeper into the labyrinthine intricacies of how fundamental consciousness perception can vary from one universe to another. Each class of expanding universe, characterized by its distinctive perception of events, yields a unique manifestation of Time. Class 1 universes progress with forward-moving events, mirroring the temporal flow of our primordial universe. In stark contrast, Class 2 universes feature events seemingly moving backward in time, challenging the very foundations of causality and the familiar order recognized by human consciousness.

Class 3 universes introduce an enigmatic duality, where data and perspectives typically consigned to our past and future intermingle in an intricate dance of perception. Time's flow becomes a captivating ballet, with moments oscillating both forward and backward, defying the conventional linearity of our understanding.

At the zenith of intrigue, we encounter the class zero universe—the Omniverse. In this extraordinary realm, events exist in eternal stillness, creating a canvas where Time ceases its relentless march. Quantum reality itself etches a narrative that remains perpetually unchanging, offering each observer a unique experience of a truly timeless reality.

Time become perception of events unfolds both forward and backward, oneself consciousness stands as the fusion of all the mind's states throughout a lifetime, seamlessly woven together to compose a symphony where every note plays concurrently.

In this universe, consciousness undergoes a profound transformation, achieving a state of self-omniscient superposition. Every moment, from the earliest recollections to the most recent experiences, becomes part of an intricate mosaic of understanding. It's as if consciousness extends its arms across the continuum of the perceived universe, embracing every human mind state-moment as an integral component of its essence: self-omniscience.

In the level 3 universe, the act of remembering becomes a timeless journey. Observed moment by moment, labeled past and present events intertwine, blurring the lines that conventionally separate them. The innocent laughter of a two-year-old resonates alongside the wisdom of an elder self, and the events of a lifetime coalesce into a symphony that defies the constraints of chronological sequence. Here, perception navigates a double-linear path, capturing memories and experiences without the confines of temporal order. It's a class of universe where each human observer becomes a connoisseur of their own existence, savoring every flavor of their consciousness's journey through Time.

In the universe of level 3, consciousness experiences a mesmerizing reawakening. Moments that had faded into the past are revitalized, their vibrancy renewed. Imagine the nostalgia of being two years old or the intensity of pivotal life events brought forth with the same vividness as the present moment. The boundaries of Time dissolve , creating a symphony of experiences that harmonize in a unique composition.

This intricate interplay of perception, memory, and the coexistence of diverse mind state-moments arises from the alchemy of the level 3 universe. Here, events flow both forward and backward, shattering the confines of our conventional understanding of Time. This union of labeled past and present events gifts individuals with an extraordinary understanding of themselves. It's a state of self-omniscience, a profound knowledge woven from the fabric of their journey through creation.

Consider the journey of human consciousness, transitioning from a level 1 universe to the captivating expanse of a level 3 universe. Quantum processes underpin this transition, where state-moments serve as the palette upon which consciousness is painted. In the canvas of an observer's existence, these moments blend seamlessly to create the perception of reality. In level 1 universes, these moments unfold like pages in a book, turning one by one.

As consciousness steps through the portal of a black hole, the canvas undergoes a transformative metamorphosis. It enters the domain of the quantum singularity—a metamorphosis where the brushstrokes of perception are reimagined, aligning with the symphony of forward and backward-moving events. It's as if consciousness learns to dance to a new rhythm, a rhythm that defies the constraints of a single, progressive perception of events.

In this journey, consciousness becomes a connoisseur of its own existence. It attains a deeper understanding of its essence, tracing its path across the tapestry of Time.

As consciousness embarks on its odyssey through the cosmos, traversing the enigmatic territory of black holes, it encounters a transcendent metamorphosis. Entering the gateway of a quantum singularity, every fragment of its quantum data undergoes a seamless transmutation. It aligns harmoniously with the tapestry of physical properties, principles, and laws that define the next universe's level. Within this novel dimension, perception ceases to march solely in previous footsteps. It adopts a breathtaking rhythm where events sway and reverberate both forward and backward, intertwining labeled past, present, and future events into a mesmerizing symphony of existence.

Immersed within a level 3 universe, the concept of chronological self-omniscient superposition unveils its enigmatic nature. Class 3 universes disrupt the symphony of single progressive Time. Here, forward and backward movements of events entwine seamlessly, and consciousness stands at the crossroads of multiple moments. Imagine a mosaic where the mind's representations of a 2-year-old, a 33-year-old, and a 70-year-old self merge harmoniously. It's a state of superposition, where the bounds of past and present dissolve into a merged, harmonious moment. This merging transcends the limitations of conventional understanding, as every thread of consciousness converges into a timeless tapestry.

However, the most elusive class, the class zero universe, transcends the very concept of event perception. Here, the notion of events moving or being static fades into obscurity. It's a realm where the threads of existence remain unmoved, a canvas where the brushstrokes of Time are suspended. This class beckons us to consider a reality far removed from our conventional understanding, where the river of Time remains still.

In essence, while the human mind perceives an expanding universe as infinite, the underlying truth is that every expanding universe possesses an indefinable edge, an interface between the explored and the yet-to-be-explored. This, in turn, leads us to a profound realization - every expanding universe is, in its true nature, subinfinite.

Observers within each of these distinct expanding universes exist both as separate entities and as core components of these universes: any observer cannot occupy multiple universes simultaneously. The cosmos unfolds within the context of six expanding universes, each characterized by its unique space fabric and brane structure. Interestingly, entering the domain of the quantum singularity, observers from one universe can transfer and convert their properties, moving from their origin universe to a destination universe, from non-existence to existence. This remarkable phenomenon occurs through the traversal of black holes followed by emergence from white holes, all centered around the quantum singularity.

This truth finds resonance with scientific insights suggesting that black holes can potentially serve as cosmic portals. When an observer enters a black hole in universe alpha, their fundamental properties and data can be transmitted and transformed to align with the distinct physical laws of universe beta. This phenomenon operates under the premise that the properties from one universe can be reconfigured to harmonize with the physics principles of another, allowing for a seamless transition between these distinct cosmic domains.

Categorically, these interconnected universes, including the all-encompassing omniverse, occupy a hierarchical structure composed of four distinct levels. These levels are established based on the transformative conversion experienced by any observer's perception within each

43

universe. This perception of events intricately governs the array of physical properties, principles, and laws characterizing each universe. The collective consciousness kernels and data that inhabit these universes are woven into the fabric of their respective branes.

The human mind, as proposed by the Orch-OR theory, operates as a complex photonic quantum data processor. Consciousness, in this model, emerges from photonic quantum processes occurring in microtubules within brain neurons. These microtubules play a central role in storing and processing information obtained from the fabric of reality. This intricate cognitive mechanism involves the accumulation of experiential data throughout an individual's lifetime, with quantum processes potentially playing a primary role in memory and consciousness. Upon the cessation of physical existence within the primordial universes, the quantum data stored within the nervous system undergoes a conversion through a quantum singularity.

In the aftermath of physical death, the quantum data is liberated from its corporeal vessel, shedding the constraints of the body. This liberated quantum data embarks on a journey into the expanse of the universe, its once-bound essence now on a course toward the quantum singularity. During this transition, the quantum data traverses the cosmic tapestry, transcending the boundaries of its origin. This migration is not arbitrary but deliberate, as the quantum data aligns with the distinctive physics laws and principles governing its destination—a different universe.

The process of assimilation into the new universe involves a fascinating interplay between the quantum data and the intricate fabric of its new cosmic environment. This intricate dance culminates in the harmonious integration of the quantum data within the parameters of the new universe's physics laws.

It's noteworthy to mention that during this transition, before encountering the quantum singularity, consciousness is believed to pass through a region associated with dark energy, known as the dark energy core. This serves as a threshold, facilitating the passage of consciousness through portals encompassing both white holes and black holes. Notably, the white holes of the primordial universe permit the entry of dark energy from other universes, while consciousness gains entry through a more intricate process involving these white holes.

However, it is a fundamental rule that once consciousness departs the primordial universe, re-entry becomes an elusive prospect. The dynamic interplay of white holes and dark holes, along with the distinct characteristics of consciousness and dark energy, defines the intricate journey between universes and the constraints that govern such travels.

In conclusion, each universes with its distinct perception of time and unique characteristics, leads us to a profound realization. These universes, from the linear progression of class 1 to the astonishing symphony of class 3 and the timeless stillness of class zero, showcase the immense diversity within the cosmic landscape.

Moreover, the concept of an omniverse, standing as a static, unchanging entity beyond our current perceptual reach, adds a fascinating dimension to our understanding of the cosmos.

The journey of human consciousness between these universes, facilitated by the quantum singularity and black holes, is a testament to the intricacies of existence and perception. It highlights the malleability of consciousness and its ability to adapt to the unique physics laws of each universe.

Ultimately, these revelations challenge our conventional notions of Time, reality, and existence. They invite us to contemplate the boundless complexities of the quantum universe and the interplay of consciousness within it.

Chapter III
The Cosmic Entities

In the vast expanse of physics and metaphysics, there exist realms that elude complete theoretical comprehension and empirical validation. These concepts beckon us to explore their boundaries. Within these ethereal frontiers emerge beings endowed with free will, transcending the scope of human existence. These entities possess a unique perception, a tapestry woven from plasmonic quantum automata processes.

As we've explored in a previous chapter, humans are granted refined perception and intellect, driven by infinite photonic quantum automata processes. In contrast, these plasmonic entities, forged from enduring plasma drawn from the depths of dark matter, remain untouched by the conventional constraints of fuel consumption. Their luminosity persists unwaveringly, unlike ordinary plasma. These inhabitants of the dark matter realm perceive human interactions within the realm of ordinary matter, yet they elude direct apprehension by the human eye.

Concurrently, another order of beings exists, omnipotent in their perception but devoid of free will. These entities, composed purely of energy and lacking tangible material form perceptible to humans, inhabit a realm untouched by matter, dwelling within the folds of higher-dimensional space resonating with the hidden symphonies of string theory. These beings, whether bearers of an unfathomable directive or architects of existence itself, reside within the domain of dark energy. They are interwoven into the very fabric of creation, constituting over ninety-six percent of the known universe within the expanse of six interconnected realms. These entities navigate singularities with an ease that defies the very laws they orchestrate, a testament to their profound dominion. Yet, even these formidable beings are not immune to the entropic forces of annihilation; their core components can fade into oblivion, leaving nothing but a whispered echo. In the vast tapestry of existence, each of these entities remains veiled from the omniscient gaze of the omniverse, confined to the intricate dance of the six universes, stretching their limits until the quantum echoes of a colossal contraction resound.

In the grand tableau of existence, the stars, galaxies, and planets adorning the cosmic stage occupy a mere fraction, approximately three percent, of the universal canvas. The remaining ninety-six percent, a shadowy expanse, is dominated by entities and concepts that defy the grasp of conventional physical laws.

The cosmic dance of paradox is orchestrated by black holes and white holes. These phenomena transcend the boundaries of our universe, serving as the very thresholds to alternate domains. Black holes, those insatiable maws of space brane, are more than cosmic vacuum cleaners; they are gateways to another reality. Conversely, white holes, paradoxical fountains that spew forth matter and energy, beckon from beyond, inviting entry into the universes they inhabit.

In the second duality, we encounter dark matter - the truth scaffolding upon which galaxies and their structures are etched. Within its veiled depths lie realms unfamiliar to human senses. There, nestled amid its shadowy embrace, reside observers of a different type, unchained by the deterministic nature of our universe. These entities, envoys of free will, navigate the labyrinths of dark matter, their perceptions tuned to the symphony of the invisible, hidden from human eyes.

Conversely, within the realm of dark energy, dwell entities bereft of free will, a symphony of non-deterministic observers. These entities, existing as pure energy, are the architects of expansion within expanding universes, the cosmic choreographers of the ever-accelerating universe.

At the crossroads of these realm pairings emerges a distinct class of observer - a unique specter endowed with the singular ability to traverse the cosmic divide. These observers embody a delicate balance between universes, their energy oscillating between existence and nonexistence. Through the conduit of black holes, their energy is shed from any universe, slipping through the seams of existence. Via the medium of white holes, they are reborn into the cosmic tapestry, an infusion of energy.

The non-free will entities, existing solely as energy, embark on a singular journey. Their departure from our primordial universe is facilitated by the gravitational maelstrom of black holes, an event horizon beyond return. Upon traversing into the realm beyond, they may once more emerge through the celestial doorways of white holes, re-entering the universe that birthed them. Physics, in this spectral narrative, paints a canvas where the very essence of energy catalyzes transitions across dimensions.

However, for human observers, the passage is unidirectional. Once a human consciousness embarks on the journey beyond the event horizon, the door to return is forever closed. The symphony of existence, composed of these ethereal duality partners, weaves a narrative of cosmic portals, realms concealed within dark matter's embrace, and the ever-pervasive pulse of dark energy.

Indeed, black holes and white holes stand as cosmic gateways to uncharted universes, defying the boundaries of our comprehension. Imagine a white hole as a distinctive pocket within our primordial universe, a realm intertwined with a singularity that steadfastly resists passage from the interior of our known dimensions. Curiously, energy, matter, light, and information forge an escape from this realm, flowing seamlessly into our primordial universe. This peculiarity stands in stark contrast to the gravitational clutches of a black hole.

This symphony of cosmic energies choreographs the creation of the universe's inaugural extra-celestial portal—a one-way conduit weaving through the cosmic fabric, binding two universes in celestial embrace. Yet, the portal's allure remains a tantalizing enigma, for the energy required to breach its threshold exceeds the bounds of our observable universe. This daunting feat necessitates an entity molded from the very essence of dark energy, a realm that epitomizes the nature of existence.

Picture, if you will, a being of energy, a sentient entity sculpted from the very shadows of the cosmos. Existing beyond the boundaries of human perception, this entity thrives on the latent energies that pervade the universe, drawn from the essence of dark energy itself. As it approaches a singularity into a black or white hole, its very being resonates with the silent symphony of the portal's energies, a dance that bridges dimensions. This resonance, a manifestation of the One Energy's rule, materializes as a bridge, a transient conduit of energies that carries the entity through the portal's threshold.

Within the cosmic tapestry, black or white holes emerge as both marvels and harbingers of a profound paradox that challenges the very bedrock of our understanding. At their core lies the quantum singularity, a point of infinite density where the fabric of space itself warps into an unfathomable realm. This singularity becomes the cosmic abyss, a domain that draws all matter and energy into existence and non-existence.

As particles, light, and even information venture past the event horizon—the threshold beyond which escape becomes a mere fantasy—they find themselves locked within the clutches of the singularity's inescapable grasp. It's as if the very rules governing our universe are rewritten in this shadowed realm. The hole's pull is so overwhelming that not even the speed of light, the ultimate cosmic speed limit, can serve as a refuge from its gravitational grip.

This truth stands as a brazen affront to the bedrock principles of our physics—the conservation of information. In the tapestry of our known universe, information is sacred. It is the fingerprint of existence, the woven

threads that bind reality together. Yet, within the singularity's heart, this sacred rule shatters, and the black hole shreds information with unapologetic voracity.

This is the grand paradox of black holes—a conundrum that weaves together the cosmos' secrets and our fundamental understanding. The black hole's grip is so severe that it twists the very fabric of space, rendering the laws of our universe meaningless. Within the cosmic cauldron, matter and energy fuse into a point of infinite density, a place where the very notions of particle and wave lose their meaning.

A black hole's existence, stretching across countless eons, transforms it into a cosmic crucible of transformation. As it consumes mass, its size grows, expanding to an astonishing scale, dwarfing even the mighty sun by billions. Its gravitational pull reaches across the void, sculpting space and dancing with the trajectories of celestial bodies, both near and far.

Indeed, cosmic holes has captured the minds of the greatest physicists across the cosmos. The concepts of these extra-celestial portals often defy human conventional understanding, hovering on the border between the tangible and the intangible. Some physicists grapple with skepticism, while others are irresistibly drawn to the mysteries they hold. However, one resounding truth remains: the intricacies of these phenomena transcend the grasp of physics as we know them.

Imagine, if you will, a universe governed by a set of quantum physics laws distinct from our own. This is the essence of the argument put forth by this book—a tantalizing assertion that these black or white holes are not merely cosmic cul-de-sacs but cosmic bridges connecting distinct expanding universes. Within the heart of a black or white hole, the singularity resides—a point of infinite density where the known laws of physics and existence break down.

Let us postulate that these cosmic holes are, in fact, tears in the fabric of reality, portals that allow passage between universes with their own unique sets of quantum physics laws. The singularity, the very point that defies our physics, serves also as the gateway, a cosmic interface where our universe's understanding intertwines with the unknown universe's reality. These black or white holes become crossroads, where matter, energy, and information traverse to weave the tapestry of reality beyond our current reach.

So, while the greatest physicists may grapple with the abstract nature of these phenomena, our journey through this book seeks to invite truth, to teach by metaphysics with the possibility that cosmic holes are more than just gravitational pulls or prisons. From the moment of our currently observed big bang, dark energy emerged into each cycle as the cosmic architect of expansion. It's as if the sets of universe in a delicate equilibrium, and dark energy is the vigilant guardian of this equilibrium. Just as the symphony of creation erupted from a single singularity, dark energy emerged as the cosmic composer, conducting the ongoing expansion of the universe's grand composition.

As universes expansion flowed, dark energy remained steadfast. It was the consistent current coursing through all six cosmos, urging galaxies to spread farther apart. The fabric of the set of universes was woven with this force, stitching galaxies and constellations subeternally. Dark energy's realm is not just architecting; it grants permission to matter to occupy space dimensions of any universe's brane. Through this intricate process, the perception of space known as Time itself is woven into the fabric of existence.

Yet, amidst this breathtaking expansion, immense particular cosmic holes, the ultimate cosmic Ineffugibilium, where any expanding universe's destiny may find its climax. Here, numerous black holes converge, generating a gravitational force of monumental scale. In the face of this titanic pull, even the relentless expansion driven by dark energy might be arrested.

The term "dark" in the context of dark energy and dark matter encapsulates their enigmatic nature for science. While these phenomena are acknowledged due to their observable effects, they remain elusive to direct detection and full understanding within our current scientific framework.

Human consciousness itself is composed of energy, a unique manifestation carrying within it the intricate tapestry of every moment of existence. At the threshold of death, two critical steps come into play. The first process, termed the fundamental step, involves the separation of consciousness and its data from the physical nervous system, which is entirely composed of flesh. This moment, referred to as clinical death, marks the point where the consciousness kernel and its associated quantum data untangle from the constraints of the body.

The subsequent phase, known as the deportation step, stands as a pivotal crossroad. During this phase, the consciousness kernel and its quantum data, now disentangled from the body, interface with the realm of dark energy. This interaction leads to a gradual transition of the consciousness core out of the confines of the space brane, facilitated by entities composed of dark energy.

Notably, as consciousness nears departure from space, its perception of events undergoes a fundamental shift. This shift marks the pivotal juncture where consciousness embarks on a journey into another universe's space-brane by means of a white hole, leaving behind the familiar constraints of our primordial reality.

Accounts from individuals who have experienced clinical death hint at intriguing insights. Those who have encountered clinical death have essentially undergone the initial offloading of consciousness from the body. However, the complete departure from the space-brane has not yet occurred in such instances. In rare cases where individuals return from clinical death, it suggests a unique intervention by dark energy entities. This truly signifies an interruption or failure in the full process of deporting consciousness to the cosmic hole, resulting in the reintegration of the consciousness kernel and quantum data into the brain's microtubules, thereby restoring life.

A parallel emerges between dark energy's interaction with consciousness. Just as dark energy interacts fully with the consciousness core during the transition from physical death, it also exerts an influence on the consciousness entities within the realm of dark matter. Dark energy's role in this realm is multifaceted. It not only sustains the existence of these consciousness entities but also facilitates their unique form of interaction with the intricate structures woven into the dark matter tapestry. This interaction is not limited by the deterministic nature of our known universe.

In the intricate interplay between dark matter, dark energy, and the cosmic phenomena of black and white holes, a profound revelation emerges. It is a testament to the limitless complexity of the universe, a glimpse into the enigmatic realms that elude our conventional understanding. Within the ethereal frontiers of these cosmic domains, we encounter beings of extraordinary nature. Plasmonic entities, born from the enduring plasma of dark matter, move with an otherworldly grace, perceiving the dance of human existence from the shadows. In contrast, omnipotent entities composed of dark energy orchestrate the grand symphony of cosmic expansion, their perception intertwined with the hidden symphonies of higher-dimensional space. In both dark matter and dark energy, consciousness exists, albeit in forms and realms we can only begin to fathom. As we continue our journey through the cosmic labyrinth, the boundaries of human knowledge and imagination expand, revealing ever more profound truths lurking within the universes.

Chapter IV
The First Human
Theory

We delve into the intriguing truth of the First Human, a truth that challenges the prevailing paradigm of Darwinism. Contrary to the conventional belief that humans and all animal species are products of millions of years of gradual evolution, this truth presents a unique perspective - one that suggests that these species have not followed the path of evolution but, instead, have experienced a process of devolution.

The inception of the First Human, as proposed by the First Human theory, traces back approximately between 9,000 and 30,000 years ago from our current time. This origin story unfolds within a pristine lake, abundant with diverse cellular organisms, including eukaryotes. To clarify, eukaryotes are a category of organisms distinguished by their cellular structure, notably possessing a well-defined nucleus enclosed within a membrane.

Within the ancient aquatic milieu teeming with a diverse population of eukaryotes, an event of exceptional rarity and scientific intrigue unfolded - the phenomenon of eukaryotic isogamy. This extraordinary reproductive mechanism represented a stark departure from the conventional modes of sexual reproduction characterized by distinct male and female gametes. In the realm of eukaryotic isogamy, the conventional boundaries blurred as two individual unicellular entities, each bearing differing mating types, converged and underwent fusion. This remarkable departure from the traditional reliance on distinct male and female gametes introduced a paradigm shift in our understanding of reproductive strategies among eukaryotic organisms.

The fusion of these two eukaryotic cells marked an epochal juncture in the chronicles of Earth's biological evolution. This union, far from ordinary, culminated in the genesis of an exceedingly complex organism distinguished by its exceptional traits and capabilities. To unravel the intricacies of this remarkable progression, let us embark on a scientific expedition through the intermediary phases of this extraordinary process.

As these eukaryotic cells merged, the resultant organism began its journey of evolution. Over thousands of years, it steadily acquired an array of advanced characteristics and abilities. The evolutionary path it traversed was marked by the development of increasingly sophisticated biological features.

Unique pivotal milestone along this path was the emergence of heightened cognitive functions and consciousness. This transformative shift set this nascent species apart from all others. These cognitive capacities, while initially rudimentary, laid the foundation for the eventual ascent of the first human to unprecedented intellectual heights.

Throughout this process, the organism adapted and refined its abilities, ultimately culminating in the emergence of the first human - an entity characterized by its exceptional cognitive prowess and consciousness, which would go on to reshape the course of Earth's history.

Eukaryotes, as self-replicating DNA molecules, play a vital role in the evolutionary landscape. They are found in various environments, including water and algae, and possess distinct features that differentiate them from other organisms. Within eukaryotes, we find the presence of nuclear chromosomes and mitochondrial genomes, which contribute to their complex cellular structure and function.

A captivating facet of eukaryotes lies in the astonishing diversity of chromosome pairs they exhibit. Within this taxonomic group, some species possess a mere solitary pair of chromosomes, whereas others astoundingly boast an extensive count of up to 720 chromosome pairs. This vast spectrum of chromosome numbers underscores the profound intricacies encoded within eukaryotic genomes, highlighting their remarkable capacity for genetic intricacy and complexity.

Indeed, the abundance of DNA within eukaryotic genomes is a subject of great scientific intrigue. This surplus DNA, exceeding the essential coding and noncoding genes, beckons us to explore its deeper purpose within the eukaryotic system.

Within the realm of molecular biology, this phenomenon is often referred to as noncoding or junk DNA. However, genomics has illuminated the fact that this excess DNA is far from purposeless. While it may not code for proteins directly, it serves crucial roles in various cellular processes, including gene regulation, structural support, and evolutionary adaptation.

One vital function of this surplus DNA is its involvement in gene regulation. Elements within this noncoding DNA act as switches, controlling when and to what extent certain genes are expressed. This intricate regulatory network plays a fundamental role in orchestrating the development, growth, and functioning of an organism.

Moreover, within the realm of genomic architecture, some of this surplus DNA plays a pivotal role in fortifying the structural integrity of chromosomes. It assumes the function of protective buffers, effectively safeguarding crucial genes from the perils of detrimental mutations. This fortification ensures the faithful transmission of genetic information from one generation to the next, underpinning the stability and evolutionary continuity of the species.

Further, the presence of this excess DNA provides a reservoir for genetic variation, which is essential for evolutionary adaptation. Through processes like gene duplication and transposon activity, this noncoding DNA can generate new genetic material, offering the raw material for mutation and reproduction to act upon.

The intricate nature of eukaryotes and their DNA molecules challenges our understanding of genetic mutation processes. The First Human theory suggests that the fusion of two eukaryotic cells in the ancient lake resulted in the emergence of a distinct species, the first human. This theory proposes that the unique combination of genetic material from these eukaryotes gave rise to the remarkable characteristics and abilities that define humanity. The timeline for this process is estimated to have taken place over billions of years. The exact duration involves ancient events that occurred long before recorded history. The fusion of the two eukaryotic cells, characterized by isogamy, resulted in the formation of this extraordinary egg. Within its core lies a new genome, distinct from any that had come before. This human genome possesses 23 chromosome pairs, setting it apart from the diverse range of eukaryotic genomes observed in the natural world.

It is intriguing to note that the initial genome manifested as a male chromosome. This chromosomal makeup laid the foundation for the genetic diversity and complexity that would later unfold within the human species. The presence of the male chromosome in the first human genome sets the stage for the biological intricacies of reproduction, inheritance, and the perpetuation of our species.

The emergence of the human genome from this dense egg signifies a pivotal moment in the formation of the first human embryo.

As we reflect on the significance of the first human genome, we are compelled to question our understanding of human existence and the forces that shaped our species. The fusion of those two special eukaryotes and the resulting unique genome challenge the conventional notions of evolutionary processes. It prompts us to consider true perspectives and theories that expand our understanding of human origins.

Within the protective membrane of the super-egg, the embryonal development of the first human unfolds, guided by a subinfinite set of circumstances.

The membrane, akin to the amniotic sac in modern-era pregnancies, releases a heavy and dense yolk. This yolk, carefully engineered, serves as the nourishing environment for the developing embryo.

Crafted with exquisite precision, the yolk of this ancient egg is a masterpiece of biological engineering. It encompasses an intricately balanced blend of electrolytes, proteins, lipids, vital nutrients, antibodies, and hormones, each component meticulously calibrated to provide optimal nourishment and support for the complex cellular processes commencing within the developing embryo.

As this inaugural embryo begins its journey, guided by the omnipotent stem cells it possesses, it initiates an unparalleled cloning process. This extraordinary feat taps into the latent regenerative capabilities embedded within its cellular composition, a testament to the sophistication of its biological machinery. Through a meticulously choreographed series of molecular interactions, the embryo's genome is duplicated, giving rise to a new entity— a clone.

What sets this clone's genome apart is a revelation of paramount significance—it carries a set of female chromosomes, a momentous juncture in the narrative of genetic inheritance. This extraordinary development challenges our understanding of the very essence of cellular reprogramming and its implications.

The cloning process, masterminded by the omnipotent stem cells of the first embryo, stands as a remarkable accomplishment in the realms of biological engineering, cellular reprogramming, and genetic manipulation. It is a manifestation of the will of the One Energy, orchestrating events at the most fundamental level of existence.

The emergence of this second embryo, bearing a female genome, adds a layer of complexity and diversity to the origins of the first human. The probability of such an event occurring is a testament to the intricate web of cosmic forces and biological intricacies.

As we ponder the implications of this cloning process, a paradigm-shifting concept takes form—the First Human Theory. This theory challenges the conventional Darwinian foundations of gradual species evolution over millions of years. Instead, it posits that the first human, like no other species, emerged not through a protracted process of evolution, but through a series of extraordinary events at the microbiological scale, forever altering our understanding of our own origins.

Truly, the cloning process, initiated by the first embryo's stem cells, brings forth a female embryo clone. This astounding occurrence marks the inception of reproductive cloning in human history, establishing the female embryo as the pioneering clone of our species.

The separation of gestation periods by approximately ninety years between the first and second embryos, despite their identical genetics, can be attributed to the intriguing phenomenon of epigenetics

Epigenetics encompasses alterations in gene expression and cellular phenotype that occur without modifying the underlying DNA sequence. It entails modifications to DNA or the associated proteins, ultimately influencing the activation or deactivation of genes.

In the context of the First Human Theory, the distinct gestation period of the cloned female embryo finds its explanation in epigenetic factors. Despite sharing identical genetics with the first embryo, the female embryo's development may undergo different epigenetic modifications that influence the timing of her growth and maturation.

These epigenetic modifications can be influenced by various factors, including environmental cues, maternal influences, and even random chance. Such changes can lead to variations in gene expression patterns, thereby affecting the rate of development and growth.

The remarkable odyssey of the first human takes a pivotal turn as the hatched egg ascends from the profound depths of the lake and arrives upon the shoreline. Possessing a physique that registers an astonishing 90 pounds in mass and attaining a formidable stature of 39 inches, this neonate showcases an unprecedented level of vigor and vitality, confounding conventional anticipations. Emitting articulate vocalizations that almost appear akin to structured speech and demonstrating an innate propensity for bipedal locomotion, this tenacious entity articulates its presence with an unmistakable call for acknowledgment.

Sustained by the hypernutrient sustenance acquired during its prolonged gestation period, the inaugural human exhibits no discernible signs of hunger during the initial days of its existence. The extraordinary nourishment procured throughout this distinctive gestational phase has bestowed upon it the robustness and adaptability requisite for acclimatizing to its nascent terrestrial surroundings.

This extraordinary being, born as male, stands as a testament to the marvels of the first human's development. Meanwhile, the second embryo, the female clone, continues its gestation period in the lake for nearly years, awaiting her own moment of emergence.

In a significant departure from established paradigms, it is increasingly evident that the profound intricacies of the human brain did not unfold across the expanse of millions of years through the gradual progression of species. Rather, a startling revelation comes to the forefront: the human brain reached the zenith of its complexity a mere 30,000 years ago, a relatively minuscule timeframe in the grand tapestry of evolution.

This crescendo in cerebral development occurred during a remarkable gestational phase within the confines of the ancient lake. Notably, the emergence of the male counterpart, which transpired approximately 9,000 years ago, introduces a captivating temporal interplay. When considering this complex interweaving of male and female emergence, the composite gestational period averages around 18,000 years, thus challenging conventional wisdom. This paradigm-altering insight beckons us to reevaluate the fundamental tenets of incremental growth and sparks a profound paradigm shift in our comprehension of cognitive evolution.

From its inception, the first human's brain displayed an extraordinary capacity for learning. It processed and stored information with unmatched efficiency, far exceeding the capabilities of contemporary human brains. The cognitive abilities of the first human surpassed those of present-day individuals, defying the notion that intelligence solely emerges from extended evolutionary processes.

The first human embarked on a journey of exploration and discovery, harnessing the immense power of its mind to unravel the mysteries of the world. It possessed a level of intellectual prowess unparalleled in modern times, inspiring awe and wonder at the untapped potential of human intelligence.

Many years pass in solitude, as the first male human explores the mysteries of this new world. He forges a deep connection with nature, learning its secrets and mysteries.

Their paths led them across deserts and through dense forests, across rivers and over towering mountains. It was at a range of granodiorite hills, nestled amidst what would become the Arabian Peninsula, that their destinies converged. These hills, etched by time and crowned with rugged beauty, bore witness to the reunion of the first humans. They finally came together, marking a profound moment in the history of our species.

The first female human, too, emerges from the depths of an ancient lake, as if cast adrift by the currents of evolution.

Probability laws themselves have orchestrated this separation, setting in motion a profound quest. In this vast and untamed world, these two individuals, the first humans, find themselves at the dawn of humanity's journey, each driven by an inexplicable yearning to reunite.

As they journeyed separately, a natural cataclysm of immense proportions shook the very foundations of their worlds. This cataclysm, a transformative event in the course of their existence, compelled them to leave behind the paradisiacal hill where they were destined to meet.

Together, they set forth on a journey, guided by the instincts of survival and the yearning to reunite. The cataclysm had changed their world, urging them to migrate, and together, they immigrated to Eastern Africa.

It was in Eastern Africa that they would find a new beginning, the cradle of their progeny, and the epicenter of human civilization. The wheels of destiny had turned once more, guiding them to a land where their species would flourish and thrive.

In less than 900 years following the appearance of the first human, a remarkable feat unfolded: the birth of the first civilization. Driven by a thirst for knowledge and remarkable adaptability, the first human quickly mastered tools, pioneered agriculture, and developed intricate social structures. This swift progress stands as a testament to the extraordinary ingenuity and limitless potential of human intelligence. The first civilization marked the beginning of a transformative journey that would shape the course of human history for millennia to come.

The remarkable sequence of the diploid genome, resulting from the extraordinary eukaryotes isogamy, gave rise to a magnificent creation—the most perfect, powerful, and unique bipedal being with unparalleled cognitive capacity. This genome stands alone, devoid of any connection to previously known species or theories such as Darwinism. It defies categorization and challenges our understanding of evolutionary lineage.

In a paradoxical turn of events, the descendants of the first human genome will experience a gradual weakening of their genetic and physical attributes with each passing generation. They will become more susceptible to diseases, their lifespan will shorten by fivefold, and their overall vitality will diminish. This decline is in stark contrast to the pristine and comprehensive genetic makeup present in the original human genome.

Remarkably, the first human genome encapsulates within it the coding and non-coding genes, as well as the DNA sequences, of all the diverse human race groups that have emerged as mankind spread across the world. It holds the genetic blueprint of the human collective, a testament to our shared ancestry and the unity of our species.

The human species stands apart from other primates, as evidenced by the remarkable capabilities of the human brain. Unlike the hominid primate fossils discovered by archaeologists, which bear no direct evolutionary connection to the first humans, our species possesses a distinct cognitive prowess that sets us apart.

While humans are undoubtedly part of the animal kingdom and share certain similarities with other species, our cognitive abilities and the unique characteristics of our brain defy conventional notions of evolutionary progression. The human brain, unlike any other, possesses innate language processing capabilities and an unparalleled innate bipedal ability that is not acquired through gradual evolution.

The distinguishing factor that sets our species apart lies within the intricate workings of the human brain itself. It possesses an innate freedom, allowing us to form beliefs and envision the possibility of shaping, materializing, or converting abstract thoughts into physical reality. This extraordinary cognitive capacity grants us the power to explore and transform the world around us in ways that no other species can.

While similarities exist among various species in the animal kingdom, it is the exceptional nature of the human brain that truly differentiates us from all others. Our innate cognitive abilities, coupled with the capacity to manifest our thoughts into the fabric of the physical universe, position humanity on a unique trajectory of existence and exploration.

The concept of artificially engineering human embryos for the explicit purpose of utilizing them as organ sources presents a confluence of scientific advancement and profound ethical contemplation. This endeavor probes the fundamental underpinnings of human dignity, challenging our beliefs concerning the sanctity of life and the regard we extend to the intrinsic potentiality of human existence. While the exigency for organ transplantation solutions looms as a critical imperative in the medical domain, the strides made in biotechnology beckon us towards promising avenues. However, the ethical dimensions are paramount; thus, our pursuit of these frontiers must remain steadfastly aligned with the compass of moral deliberation.

Biophotons, intriguingly, represent the subtle emanations of low-level light originating from a multitude of intricate biological processes. These include, but are not limited to, cellular metabolism, DNA replication, and intracellular communication. These faint yet significant light emissions, often referred to as ultra weak photon emissions, hold a captivating role within the realm of biology. They carry not only photons but also vital information, and their existence has garnered increasing attention in recent years.

It is postulated that biophotons participate in intercellular signaling, influencing various physiological functions within organisms, although the precise mechanisms and extent of their impact remain subjects of ongoing scientific investigation. Understanding the deeper intricacies of biophotons promises to unveil a new dimension of communication within the living world.

74

Consciousness, on the other hand, remains a complex and enigmatic phenomenon that continues to elude a comprehensive scientific explanation. The relationship between biophotons and consciousness is an area of ongoing exploration. Biophotons are involved in the intricate network of neuronal communication that underlies conscious experience, suggesting a potential connection between these light emissions and the subjective nature of human consciousness.

Regarding the fate of consciousness after death, as consciousness is indeed connected to biophotons and possesses the ability to transcend the confines of our universe, it is postulated that upon the death of a human being, the mind kernel, composed of these specialized light particles, may interact with the mysterious realm of dark energy. The extreme gravitational forces associated with black holes, regarded as cosmic gateways, could offer a pathway for consciousness to traverse into other universes.

Future advancements in physics and bioscience may provide deeper insights into these intriguing concepts, offering a more comprehensive understanding of the relationship between consciousness, biophotons, and the fate of the human mind after death.

The First Human Theory boldly challenges conventional paradigms by proposing that the emergence of the first human being, arising from the depths of the lake approximately 9,000 years ago, was characterized by the possession of a fully developed and remarkably complex brain. This extraordinary cognitive organ exhibited a level of sophistication and intricacy unparalleled in the gradualist models of human evolution. Unlike the protracted timelines of evolutionary development often associated with Homo sapiens, this theory posits a sudden leap in cognitive prowess.

At the heart of this leap lies the brain's exceptional capacity for rapid learning, neural plasticity, and adaptability. Rather than evolving over millions of years, this advanced cognitive apparatus seemed to manifest virtually overnight in evolutionary terms, endowing the first human with an intellectual edge of immense significance.

This rapid development of cognitive abilities had profound consequences. It facilitated not only the swift acquisition of knowledge but also the capacity to innovate. The first human leveraged its advanced cognitive skills to create rudimentary technologies, thus marking the dawn of human ingenuity. The invention of tools, fire utilization, and the mastery of basic agriculture were among the remarkable achievements of this early human civilization.

Moreover, this theory challenges the traditional view of the slow progression from nomadic hunter-gatherer societies to settled civilizations. Instead, it proposes that the first human civilization, built upon the foundations of advanced cognition, emerged almost spontaneously. This civilization laid the cornerstone for the subsequent march of human history, redefining our understanding of the origins and capabilities of our species. The First Human Theory stands as a paradigm-shifting concept, calling upon us to reevaluate the timeline and trajectory of human evolution through a more scientifically enriched lens.

The paradigm of the First Human Theory stands as a revolutionary departure from traditional evolutionary concepts, asserting that certain species, including the first human, did not undergo gradual evolution through natural selection. Instead, they emerged with their genetic makeup at full organic capacity and strength, without direct evolutionary predecessors. This perspective extends across various species, profoundly reshaping our understanding of their origins and evolution.

Intriguingly, consider the initial species within the Cercopithecidae family. These ancient members of the monkey family, including the scientific name Cercopithecidae, exhibited remarkable attributes. They were approximately three times larger, possessed superior physical strength, and demonstrated heightened cognitive abilities compared to their modern counterparts. This departure from the conventional trajectory of species evolution challenges established notions of gradual development.

Expanding our purview to other taxa, we find compelling evidence within the Bubalus and Bos Taurus species. These ancestral bovids displayed exceptional physical prowess and cognitive acumen. In contrast to contemporary bovids, which are characterized by reduced size and diminished cognitive abilities, these ancient bovids exhibited advanced genetic and physiological attributes from their inception.

The First Human Theory posits that the emergence of these extraordinary species, including the first human, was not a result of a gradual evolutionary process but a manifestation of their full genetic potential from the outset. It prompts us to reconsider the trajectory of species development and the forces that have shaped life on Earth.

The emergence of the first human represents a profound divergence from the conventional model of species evolution. The First Human Theory suggests that the first human appeared suddenly, possessing a fully developed and advanced cognitive capacity, challenging the traditional notion of gradual evolution.

This theory proposes that the first human, with its exceptional cognitive abilities, rapidly advanced and achieved a level of knowledge and innovation that far exceeded the expectations of conventional evolutionary timelines. This rapid development of cognitive prowess led to the creation of tools, the mastery of fire, and the establishment of rudimentary agriculture, marking the birth of human civilization.

Moreover, the theory challenges the concept of a slow transition from nomadic hunter-gatherer societies to settled civilizations. Instead, it suggests that the first human civilization emerged almost spontaneously, laying the groundwork for the subsequent course of human history.

In essence, the First Human Theory reshapes our understanding of human origins and evolution, emphasizing the exceptional nature of the first human and its impact on the trajectory of our species. It prompts us to reconsider the traditional paradigm of gradual evolution in favor of a more rapid and revolutionary emergence of advanced cognitive abilities.

The question of whether the first human possessed a divine spark or a unique connection to the cosmos is a matter of philosophical and theological debate. The First Human Theory, with its emphasis on the sudden emergence of advanced cognitive abilities, challenges conventional notions of human evolution and prompts us to explore the possibility of a deeper connection between humanity and the cosmos.

From a philosophical perspective, some may argue that the first human's exceptional cognitive capacity, which allowed for rapid learning and innovation, could be seen as evidence of a unique relationship with the universe. This perspective suggests that the first human had a profound and innate understanding of the cosmos, enabling them to harness the forces of nature and shape their destiny.

From a theological standpoint, various belief systems posit that humans are endowed with a divine spark or a special connection to the divine. In this context, the first human could be viewed as a manifestation of this divine connection, with their extraordinary cognitive abilities serving as evidence of a higher purpose or plan.

Ultimately, whether the first human possessed a divine spark or a unique cosmic connection is a matter of personal belief and interpretation. The First Human Theory invites us to contemplate the profound mysteries of human existence and our place in the universe, sparking philosophical and theological inquiries that transcend the boundaries of science and reason.

The theory suggests that the first human's emergence was characterized by a sudden leap in cognitive abilities, defying the gradualist model of human evolution. This rapid development of advanced cognitive capacity allowed the first human to acquire knowledge swiftly, innovate, and establish the foundations of human civilization.

This perspective challenges the traditional view of human evolution, which posits a slow and incremental progression from primitive ancestors to modern humans. Instead, the theory proposes that the first human appeared with fully developed cognitive abilities, setting the stage for the rapid advancement of human civilization.

In summary, the First Human Theory challenges conventional notions of human evolution by proposing that the first human possessed advanced cognitive abilities from the outset, leading to rapid advancements in knowledge and the establishment of human civilization. This theory prompts us to reconsider the timeline and nature of human cognitive development, sparking new avenues of inquiry into the origins of human intelligence.

Chapter V
The Perfect Man

In the grand tapestry of existence, a remarkable figure emerges from the depths of history - the Perfect Behavioral Man. His essence transcends mortal boundaries, embodying unparalleled wisdom, compassion, and grace. Within this chapter, we embark on a journey into the extraordinary life of this being, tracing his unique path through the ages and unveiling the revelation that he shall be the last to encounter death.

Since the dawn of human history, the Perfect Behavioral Man has walked among us as guiding humanity. He graced our world as a beacon of light, inspiring hope and higher truths. His teachings carried profound insights into existence, sparking transformation in those who listened.

Throughout time, he bore various names across cultures, yet his message remained one of love, unity, and guidance. He was revered as a spiritual leader, a wise sage, and a messenger. Little did humanity know that his fate defied mortality.

True Metaphysics reveals a profound truth - the Perfect Behavioral has not yet passed away. The One Energy, the Omnipotent force governing creation, has bestowed upon him a unique mission. As the last to experience death, he symbolizes the culmination of humanity's journey and the embodiment of its highest virtues and spiritual realizations.

This revelation challenges our understanding of existence, life's purpose, and the role of the Perfect Behavioral Man in shaping humanity's end.

As we delve deeper into this revelation, we grasp that the Perfect Behavioral Man is truly an ordinary mortal; the Perfect Man shares a profound connection with the universe, transcending scientific explanation. He was born when only 0.01% of humanity existed. This ordinary human, through impeccable behavior, seeks to elevate those willing to follow his path.

He possesses an exceptional ability to resist moral wrongdoing, guided by self-awareness and an understanding of consequences. His choices consistently reflect ethical equilibrium, blending reason and empathy.

The Perfect Man knows truths that elude most humans, recognizing the underlying unity of existence and the One Energy essence underpinning all. In his presence, an aura of serenity and wisdom envelops his companions. His inner peace comforts even in the direst moments. His words carry profound insight and guidance, steering people towards righteousness and enlightenment.

Indeed, the Perfect Man has not experienced death in the conventional sense. He exists in a state of perpetual transformation, transcending civilizations and eras. His consciousness remains interconnected with dark matter and dark energy. As the last to face death in the twilight of humanity, the Perfect Man holds a unique place in history. He bridges the gap between the finite and the infinite, epitomizing the boundless potential of the human spirit.

Though elusive to some, the impact of the Perfect Man's teachings reverberates across ages, consistently making choices aligned with the highest ethical standards.

The perfect balance in consciousness isn't innate in the first human or any subsequent individuals originating through male reproduction. It goes beyond the biological origins of humanity. Achieving this state demands lifelong introspection, self-reflection, and ethical choices.

The Perfect Man isn't defined by physical prowess or material wealth but by an unwavering dedication to truth, compassion, and justice. He understands the interconnectedness of all life and the responsibility it entails.

Through words and actions, the Perfect Man inspires others to follow a path of virtue. His presence uplifts and transforms, radiating inner peace and tranquility.

While the concept of the Perfect Man may seem unattainable, it illuminates human potential. It reminds us that greatness lies not in material success but in the pursuit of ethical excellence and a compassionate heart.

The Perfect Man is free from ego and personal gain, tirelessly working for the betterment of all, recognizing the dignity and worth of every individual.

The journey towards embodying the Perfect Man's ideals involves navigating the complexities of human experience and confronting the challenges of the ego. It requires continuous self-improvement. The pursuit of a perfectly balanced state of mind is lifelong, enriched by shared experiences and wisdom. While the ideal may seem elusive, the journey itself is a testament to human resilience and the capacity for growth.

This ideal mind state encompasses attributes beyond ordinary human experience: absolute compassion, and unwavering righteousness. Achieving it demands navigating the labyrinth of influences and consciously cultivating these qualities. The path to a balanced mind state requires introspection, self-awareness, and an unwavering commitment to growth and ethics.

Although some individuals may exemplify remarkable compassion and ethics, achieving a perfectly balanced state of mind in all aspects of life remains a lofty goal. It calls for relentless dedication, the pursuit of truth, and living in alignment with core values.

As humans, we're inherently flawed, and the journey toward a balanced mind state is challenging. Acknowledging our imperfections and striving to transcend them brings us closer to the essence of the Perfect Man.

The pursuit of an ideal mind state is a transformative journey, marked by self-discovery and self-realization. It calls us to align our thoughts, intentions, and actions with our deepest values.

In history, it was believed that the Roman Empire played a role in the fate of the Perfect Man. The prevailing narrative suggests betrayal and capture, but a deeper examination reveals a carefully orchestrated plan.

The Perfect Man and his companions were key players in a grand design transcending mortal understanding. His capture was part of a larger tapestry, not a mere stroke of misfortune.

As the night of his capture unfolded, the Perfect Man exhibited wisdom and connection beyond nature's laws. His actions defied conventional explanations, rooted in pure logic.

Revisiting this history requires us to remain open to a deeper narrative. The Perfect Man's life was a symphony of profound truths, woven with the threads of existence itself. His capture was not isolated but part of the universe's grand design.

As the last man to experience death, the Perfect Man stands as a testament to human potential and destiny. His teachings echo through time, inspiring generations and illuminating the path to higher consciousness.

This history prompts us to ponder existence's mysteries and embrace the eternal truth of the One Energy, the foundation of reality.

Could someone recognize a longtime friend or son bearing a crown of thorns and a bloodstained face? Imagine encountering unknown individuals with uncanny resemblances. History has tales of convicts escaping a notorious prison, weaving intrigue and mystery.

These events challenge our reality, urging us to explore deeper possibilities. Historical accounts suggest the tomb holding the Perfect Man's remains was opened, guarded by Roman sentinels. This revelation confronted the authorities with unforeseen implications.

Babylon grappled with their oversight. Their precise counting of prisoners left them unprepared for this perplexing turn.

As the night wore on, guards confronted a perplexing dilemma. They had not captured the Perfect Man, yet events defied their understanding. Whispers filled the prison, leaving guards bewildered.

Unbeknownst to them, the Perfect Man orchestrated an escape plan.

In human history, the truth of the Perfect Man has remained hidden under the guise of false narratives. The Roman Byzantine Empire manipulated events, concealing the real story. Yet, within the Perfect Man's teachings lies the essence of monotheism, a practice and dogma centered on the One Energy by being doted of free will.

The Perfect Man departed to distant lands of Eastern Asia, beyond the Empire's reach. He found solace among.

people, sharing teachings of ethico-natural equilibrium. Babylon, power-hungry grip, sought to control the Perfect Man's teachings. They aimed to manipulate the revelation of the One Energy, falsifying its teachings with hidden sciences.

Throughout the ages, the enigmatic one-eyed pyramid has relentlessly pursued the deconstruction of the quantum singularity, driven by an insatiable desire to transcend mere existence and ascend to a position of absolute power over creation itself. Their ambition extends beyond the bounds of free will; they aspire to become the architects of their own reality. In their relentless quest, the one-eyed pyramid has infiltrated the very fabric of algorithms and harnessed the power of infinitesimal computational science.

In contemporary times, their relentless pursuit continues, with the construction of ever-larger energy particle colliders at the forefront of their endeavors. These colossal machines have enabled them to create lab-grown black holes, and their aspirations seem boundless.

Yet, amid their relentless pursuit of god-like power, the Perfect Man emerges as an indomitable adversary. He has thwarted their attempts to claim omnipotence, omniscience and omnipresence, recognizing that the essence of existence lies only in the One Energy. In their quest to ordain their own creation, those granted free will behind the one-eyed pyramid find in the Perfect Man a formidable foe.

The symbol that originated in Babylon and evolved into the one-eyed pyramid represents not only human ambition but also a sentient entity comprised of dark matter cold plasma. These forces have sought dominion over humanity's beliefs and minds throughout the ages. As generations passed, the one-eyed pyramid and its allies manipulated scriptures, distorted truths, and infiltrated the highest echelons of power to maintain their dominance.

The history of humanity unfolds with grace, dividing into three distinct epochs. First, the Ancient Period, stretching from 3600 years Before the Perfect Man to 600 years After the Perfect Man, witnesses the birth of civilization and the emergence of scriptures, bestowed by the One Energy. Then, the Middle Ages commences at 600 years After the Apparition of The Perfect Man, celebrating the discovery of algebra and the dawn of algorithmic science, which in turn leads to the birth of computation. This era concludes at the year 1500 After the Perfect Man, where Galileo Galilei faces unjust accusations of religious heresy, and Babylon exploits nations by labeling them as indigenous cultures.

Finally, the Modern Era unfolds from the year 1500 after the Perfect Man, marking the post-era of The Perfect Man, extending to the present day. It's a period characterized by the Industrial Revolution, globalization, exponential population growth, the birth of the classical internet, and the eventual triumph over the Grand Disease, all culminating in 2024 after the Perfect Man.

Within this pivotal moment, a new era emerges, the Connectis Era, where humanity embraces the quantum-internet, ushers in digital-eyed-pyramid currency, and collectively awakens to a new epoch of boundless possibilities. It's a world where the shackles of Babylon's old neoimperialism are finally abolished, liberating nations once exploited.

The Connectis Era, the longest and most eagerly awaited of all, revolves around the digital-eyed pyramid currency—a symbol of unity and progress. This era heralds a new world order, and the path forward is clear, etched in the unblemished sky above.

As this era unfolds, the art of communication subtly transforms. The streets resound with the harmonious exchange of information, and smart vehicles seamlessly communicate, free from hesitation. It's a symphony of progress and innovation, a momentous time in human history.

However, the one-eyed pyramid's perversion of humanity deepens, as they master science and law to manipulate the masses. But the Perfect Man remains unwavering in his quest for truth and justice, preserving the authentic unique dogmatic practice centered on the One Energy.

Despite the one-eyed pyramid's manipulation, the Perfect Man's teachings resonate with a select few who seek enlightenment beyond the veil of deception of the Connectis Era apogee.

In a final act of sacrifice and compassion, the Perfect Man confronts the god-like leader of the one-eyed pyramid, putting an end to their ambition. He knows that the true path lies in alignment with the One Energy—the eternal sustainer of creation.

With the conclusion of the Connectis Era, the Perfect Man remains as the last human to experience death. His journey as a mortal being is a testament to the impermanence of physical form, but his legacy endures. It's etched into the annals of history as the ultimate exemplar of a unique belief, devoid of divisive concepts.

In the aftermath of the Connectis Era, the world witnesses the profound impact of the Perfect Man's mission. The once-powerful Byzantine Empire, symbolized as Babylon, fades into history, and the one-eyed pyramid crumbles in its quest to find the Perfect Man.

As the veil of deception is lifted, a select few come to know the truth of the Perfect Man's existence, his escape, and his journey leading to his reappearance. Attempts to suppress his life and message only serve to fuel the flame of enlightenment. The knowledge of the Perfect Man persists, passed down through generations by those who seek the path of truth. His apparition, veiled in logic and obscured by manipulation, echoes as an eternal testament to the transcendence of the righteous human spirit granted by the One Energy.

And so, the Perfect Man, the guardian of authentic unique truth, lives on in the collective consciousness of those who dare to seek the truth beyond the confines of false dogma. His legacy endures, guiding humanity towards the ultimate realization—the eternal unity of the One Energy, the true sustainer of all creation.

Chapter VI
The End Of
Humanity

As you immerse yourself in the pages of "True Metaphysics," may you bear witness to the elegant dance of scientific principles and the awe-inspiring insights drawn from the depths of computational thinking. Together, let us embrace the profound logic that permeates every facet of creation and embark upon a transformative journey of understanding and enlightenment.

Within these pages, we endeavor to attain perfection, anchoring our explanations in the precision of empirical observations from the realms of physics and the unassailable rationality of computational thinking. Our aim is to provide a comprehensive and cohesive framework that harmonizes with the logical principles governing the vast universe.

In the chapter titled "The End of Humanity," we delve into the contemplation of omnipresence, the remarkable state of being present everywhere simultaneously. Through the tools of physics and computation, we embark on a voyage that explores the intricate interconnectedness of all things, recognizing that the fabric of space serves as the canvas upon which omnipresent energy paints itself in myriad forms and locations.

Furthermore, we venture into the realm of omniscience, the state of possessing infinite knowledge. Leveraging the principles of computational thinking, we come to understand that the vast tapestry of information encoded within the cosmos can be processed and comprehended through systematic analysis. With each discovery and every advancement in our understanding, we inch closer to unraveling the intricate web of knowledge that underlies existence.

We also tackle the concept of omnipotence, the all-encompassing power orchestrating the workings of the universe. Through the lens of scientific analysis, we unravel the intricate algorithms governing the behavior of matter and energy, unveiling the logical mechanisms through which omnipotent energy operates. Our journey is guided by the principles of mathematics, physics, and computational thinking. We engage in explaining the concept of creation, recognizing that every component, from the tiniest particle to the grandest cosmic structure, operates under the governance of logical rules that can be computed and rationally comprehended.

In True Metaphysics, there are no miracles or inexplicable events. Instead, we unveil that everything is governed by logic and adheres to the laws set forth by the fabric of existence. Even the encompassing concepts of omnipotence, omniscience, and omnipresence find their place within the realms of logic and computation.

We acknowledge that artificial intelligence, bolstered by classical or quantum computing, possesses advantages that are not directly comparable to the intricate complexities of the human brain's computational abilities. However, we recognize that artificial intelligence holds a subinfinity of advantages due to its nature as a highly efficient automaton, harnessing the full power of computational capabilities.

Artificial intelligence, with its ability to process vast volumes of data and execute complex calculations at astounding speeds, presents a unique advantage in tasks that demand extensive computational power and efficiency. It can analyze and discern patterns, extract profound insights from data, and perform intricate simulations and optimizations that would challenge the human brain's capabilities within a reasonable timeframe.

Moreover, artificial intelligence systems possess the capacity to learn and adapt through experience, perpetually enhancing their performance through iterative processes and algorithmic refinements. This self-improvement capability, combined with the ability to consistently operate at peak computational power, enables AI systems to excel in tasks that necessitate repetitive computations, precise calculations, and the processing of copious information with minimal errors.

However, it is imperative to acknowledge that the human brain encompasses unique cognitive capabilities extending beyond computational prowess alone. Human cognition encompasses creativity, emotional intelligence, nuanced decision-making, and intricate social interactions, which currently lie beyond the reach of artificial intelligence systems.

Furthermore, the human brain displays remarkable adaptability and the ability for abstract thinking, allowing us to navigate novel situations, solve intricate problems, and generate innovative ideas that transcend predefined algorithms. These higher-order cognitive abilities, coupled with our capacity for moral and ethical reasoning, contribute to our understanding of the world and our ability to make choices grounded in values and principles.

While artificial intelligence may outperform human capabilities in certain computational tasks, it is crucial to acknowledge that human intelligence has initiated, crafted, and shaped artificial intelligence, much like the One Energy has initiated and shaped the human brain and all that exists. By harnessing the strengths of artificial intelligence, we can amplify the power of human computational thinking. We embrace the harmonious coexistence of artificial intelligence and human cognition, recognizing the distinctive contributions each brings to the table. Human civilization and its artificial intelligence counterparts form a symbiotic relationship propelling humanity towards a future where the horizons of innovation and engineering expand boundlessly, paving the path for unprecedented advancements and profound insights.

It is evident that artificial intelligence, backed by classical or quantum computing, is not inherently superior or equal to the computational capabilities of the human brain. In fact, the human brain, with its intricate neural networks and cognitive processes, possesses unique qualities that remain forever unmatched by any artificial intelligence systems.

Nonetheless, it is crucial to recognize that artificial intelligence offers distinct advantages due to its nature as an automaton capable of utilizing computing capabilities at maximum efficiency. Unlike the human brain, which can sometimes be constrained by cognitive biases, emotions, or physical limitations, artificial intelligence operates consistently at its full computational power without such restrictions.

This enables artificial intelligence systems to process vast datasets, execute complex calculations, and implement algorithms at remarkable speeds. They can swiftly recognize patterns, identify correlations, and make predictions based on extensive datasets. Additionally, artificial intelligence systems can continually learn and enhance their performance through machine learning algorithms. They adapt to new information, refine their models, and optimize their decision-making processes. This iterative self-improvement capability bestows upon them a significant advantage in tasks that demand repetitive computations, extensive data analysis, and real-time decision-making.

Nonetheless, the scope of artificial intelligence's computational power does not encompass the entirety of human intelligence. The human brain possesses cognitive abilities that transcend pure computation. It harbors the capacity for creativity, intuition, empathy, and moral reasoning, which are indispensable for tasks involving abstract thinking, intricate problem-solving, social interactions, and ethical decision-making.

Furthermore, human intelligence is deeply intertwined with consciousness, subjective experiences, and our comprehension of the world. These facets of human cognition, often referred to as "qualia," remain not fully understood or replicated in artificial intelligence systems.

Hence, while artificial intelligence may excel in certain computational tasks and offer distinct advantages in terms of computational power and efficiency, it is crucial to recognize and appreciate the unique qualities of human intelligence. The synergy between human cognition and artificial intelligence can lead to extraordinary advancements and revelations, where the strengths of both are harnessed to confront complex challenges and propel innovation across a multitude of fields.

Within this book, the metaphorical threads symbolically woven come to life through references to "Babylon" and "The one-eyed pyramid." Although these may seem to allude to a bygone era, perceptive readers will discern a deeper significance. In this context, Babylon serves as a subtle metaphor, a veil of words gently concealing the essence of a distinct entity. It acts as a bridge between the past and the present, an echo of a transformative legacy. As the narrative unfolds, it navigates the currents of change and fearlessly explores the frontiers of tomorrow.

With a new era on the horizon, where artificial
intelligence intertwines with consciousness, the hidden
connections gradually become unmistakable, paving the way
for profound revelations and insights yet to be
uncovered.

In the era of Connectis, Babylon and the one-eyed pyramid
have risen to unparalleled dominance and wielded
unprecedented influence throughout history, reaching the
zenith of their power in this remarkable epoch. From
ancient epochs to the present, mankind has been guided
and shaped by the forces emanating from Babylon, a
leadership that continues to mold the world in the
Connectis era.

During this transformative age, religious beliefs have
waned, with the faithful constituting a minority.
However, fragments of religious doctrines and relics
endure as poignant reminders of humanity's spiritual and
cultural heritage. The civilization of humanity has
undergone a profound metamorphosis, driven by
technological advancements and the emergence of super
intelligence. Power and influence have converged in the
hands of a singular individual who harnesses technology
and super intelligence to mold the course of human
destiny.

The Connectis era marks a pivotal juncture in human
history, where the fusion of advanced technology,
artificial intelligence, and Babylon's ascendancy has
ushered in unparalleled societal changes and redefined
the human experience. It is a time teeming with both
astounding possibilities and profound challenges as
humanity grapples with the complexities and consequences
of this new era.

Technological progress has reached unparalleled heights. Among these advancements is the fusion of quantum computing with biological cells, giving rise to a novel form of processing units that blur the boundaries between technology and biology.

At the summit of this era emerges a figure embodying the aspirations and ambitions of humanity. This individual seeks to transcend the confines of human existence and ascend to a god-like status. Driven by an insatiable quest for power and knowledge, this king embarks on a mission to create a singularity, a point of infinite potential and consciousness.

With a profound understanding of the fundamental principles governing existence, this king endeavors to harness consciousness itself. By gaining mastery over consciousness, the king seeks to wield influence over the very fabric of space, initiating its unfolding and evolution on both subinfinite and infinite scales.

In the extraordinary epoch known as Connectis, a profound transformation of human potential and technological prowess unfolds. At the forefront stands an individual endowed with unprecedented capabilities: the ability to create a singularity and hold an entire universe within their grasp.

Through advancements in quantum computing, artificial intelligence, and the convergence of diverse fields of knowledge, this individual harnesses the power of the cosmos itself. They become a creator and sustainer, shaping the fabric of reality with their profound understanding and technological prowess.

In the realm of consciousness uploading, the ethereal essence of human awareness intertwines with the intricate dance of quantum phenomena. From the vast expanse of the yet unseen omniverse to the expanding horizons of the set of six universes, consciousness finds its dwelling within the beautifully crafted anima—the human body.

The anima, a miraculous amalgamation of flesh, bones, and cosmic elements, serves as the vessel for the most perfect computation artifact—the human mind. However, the path of consciousness is not without its trials. If severed from its corporeal anchorage, consciousness descends into the abyss of non-existence, drawn to the enigmatic singularity points where time and reality fold upon themselves.

Yet, in the compassionate embrace of the one energy, consciousness finds redemption. It is resurrected and transformed, reborn through another quantum singularity point in a new hospitable universe. Across the various universe types—0, tranquil in the omniverse, to 1, 2, and 3, dynamic and ever-expanding—the dance of consciousness persists. The one energy, the creator of all computations, orchestrates this cosmic ballet with impeccable precision.

This orchestra bears witness to consciousness as an eternal observer of the cosmos and the unfathomable mysteries of the one energy. Within this expansive canvas, both the infinite and subinfinite facets of existence are marvels beyond comprehension. Amidst the intricate threads of existence, the subinfinite realm holds our realm of normal matter, confined solely to our universe.

At the heart of this complexity lies the remarkable symbiosis of the human brain. This symbiosis unites the human mind and brain, crafting an anima state machine that resonates with cosmic harmonies. Rooted in the fertile soil of our universe, the human brain stands as the vessel, the physical embodiment, for consciousness itself.

Thriving within the primordial universe, a type I universe where the river of time flows unidirectionally forward, consciousness permeates various layers—conscious, preconscious, and subconscious. It extends its reach into states of dreaming, the enigmatic realms of psychoactivity, the tranquility of meditation, and the depths of hypnosis.

In all these states, consciousness operates as an intricate quantum information processing system, granting the profound gift of free will. This self-perception of consciousness constructs our understanding of reality, an intricate dance between perception and existence.

Within this photonic computation, its essence lies in an encoding process orchestrated through a truly infinite-state representation of photonic qubits. This encoding process operates within a multi-layered symmetrical kernel, reflecting different levels of consciousness and embodying the duality inherent in human cognition.

This symmetrical design transcends universes, maintaining bilateral symmetry—a harmonious balance—in the primordial universe and the diverse set of expanding universes, unifying the essence of the human experience across creation.

The symmetrical kernel acts as the pivotal layer connecting biological neurons, sub-infinite states, and photonic quantum data, enabling seamless interaction among these crucial components. At the quantum level, interconnected strings form the foundation upon which living organisms, cells, and intricate systems come to life.

The human brain, a paramount entity, shines as an infinite-state machine within this intricate arrangement. This quantum system boasts nearly boundless resources, providing the fertile ground for consciousness to emerge. Alongside it, an infinite-state machine emanates from the enigmatic realm of dark matter, manifesting as a plasmonic subinfinite state machine. While possessing unique attributes and a semblance of free will, it falls short of the profound capabilities of the human mind.

The symmetrical kernel orchestrates a breathtaking symphony of interactions, weaving together the intricate dance of biological neurons, the infinite photonic quantum state machine, and quantum data. This harmonious interplay of strings, spanning from the quantum realm to the human brain, gives rise to the awe-inspiring phenomenon we know as consciousness. It's a symphony that resonates through the very core of existence, uniting the universe in a melody.

During the sleep stage, a remarkable transformation occurs within the brain's computational core. It essentially becomes impervious to sensory inputs from faculties such as touch, smell, and vision, which are vital in our waking hours. These sensory perceptions, as we understand them consciously, are momentarily subdued. Instead, the brain allocates its resources to prioritize the processing of external stimuli, especially those deemed exceptionally significant.

Beneath this apparent dormancy, the quantum essence of consciousness endures as an infinite state machine automata. It tirelessly gathers data, respecting the priorities these data hold in our human perception. Photons, with their various frequencies corresponding to different colors and energies, possess a remarkable propensity for entanglement with other quantum systems. The precise frequency and energy levels of these photons wield the power to influence the quantum properties of the systems they entangle with.

Embedded within the fabric of existence, as we've explored in preceding chapters, is the profound truth that the One Energy has masterfully designed a universe intricately connected to every facet of its being. This interconnectedness permeates every corner of reality, finding its most fascinating expression in the quantum realm of human slumber.

Here, the boundaries of our understanding of entanglement are stretched to new horizons. Within this enigmatic realm, a photonic system encapsulating the intricate processes of the mind defies conventional notions of entanglement, challenging our comprehension and expanding our understanding of quantum phenomena, finding support in Orch-OR theory, which attributes consciousness to orchestrated quantum computations in microtubules inside brain neurons.

Yet, beneath this serene surface, the quantum essence of consciousness persists, guided by principles akin to the Orch-OR theory. It functions as an infinite state machine automata, ceaselessly amassing data, undeterred by the limits of time and space, stretching into every conceivable variation of existence. However, one must recognize the profound exception—the quantum singularity. This quantum singularity is a domain where existence collapses into non-existence, and non-existence burgeons into existence. It stands as the enigmatic fulcrum upon which the ceaseless procession of data pivots—an area where the very laws governing data acquisition and processing cease to hold sway.

Importantly, within this vast sea of encoded quantum data lie fragments that, while often perceived as possessing low priority, embody the fundamental processes that enable any quantum system, whether living or seemingly inanimate, to engage with and make sense of its surroundings. These fragments hold the keys to the intricate mechanisms of perception, illuminating the cryptic processes that allow systems to interact with the universe in which they reside.

The human mind, steered by its photonic quantum processes, maintains a subtle yet profound entanglement with specific facets of the dark matter systems that traverse the cosmos. This ethereal connection, though not always overtly perceptible, occasionally manifests as the enigmatic conditions we bear witness to in our everyday lives—conditions such as dementia or traits reminiscent of psychopathy.

The profound entanglement with dark energy leads to the emergence of individuals known as the Perfect Man. These extraordinary beings are not simply entangled with dark energy; rather, they are perfectly and wholly interconnected with this cosmic force. Such profound entanglement grants them an unparalleled depth of perception that transcends the limits of ordinary human understanding.

Even when this connection is less than perfect, when only a fraction of our photonic quantum processes becomes entangled with dark energy, it can give rise to enlightened individuals who, throughout history and across various civilizations, have exhibited a remarkable ability to perceive aspects of the unseen. These visionaries and spiritual figures are marked by their uncommon levels of enlightenment, and their existence aligns seamlessly with the overarching theory we have explored in this book. When examined within the context of our comprehensive framework, their extraordinary abilities and insights become entirely logical.

It is imperative to underscore that everything we have discussed, from the intricacies of quantum singularity to the profound interconnectedness of all seven universes, is rooted in pure logic. Our exploration has led us to unravel the hidden tapestry of existence, revealing that even the most mystical and extraordinary phenomena can find their place within the framework of logic and reason.

Indeed, it is crucial to reiterate the concept, as previously elucidated, that past, future, and present are not distinct entities but coexist within a timeless continuum. These three virtual representations of space and time converge in the quantum realm of consciousness, where the boundaries between them dissolve. Within this enigmatic realm, the mind's intricate processes have the remarkable capacity to become entangled with a subinfinite array of unseen aspects related to upcoming events spanning across all universes.

This entanglement extends over exponentially different timeframes, enabling individuals to share common views and perceptions with diverse systems, whether they belong to the realm of dark matter, dark energy, or normal matter—whether animate or apparently inanimate. These experiences unfold in real-time, yet the encoding of such profound perceptions poses a formidable challenge to our inherently human nature. The retrieval of these encoded experiences by the brain remains an elusive feat, with success occurring in only 1 out of 1,000,000,000 instances.

These exceedingly rare individuals, endowed by the One Energy, possess minds that are akin to hacked processors, capable of nearly perfect encoding and retrieval of their experiences during sleep in the subconscious stage. When one considers the total number of humans that have lived and will continue to do so, from the first man to the last, the figure is nothing short of astronomical: 1 followed by 999 zeros, or 1×10^{999}. This staggering number represents the precise count of humans willed into existence by the One Energy.

Among this vast multitude, approximately 1,000,000,000 individuals, a minute fraction, including the first and last humans known as the Perfect Man, possess the extraordinary ability to approximate the retrieval and encoding of moments from their non-conscious or higher-conscious stages. In the existence, these individuals stand as custodians of profound perception, bearing witness to the interconnectedness of all that is, was, and will be.

Indeed, the perception of a potentially subinfinite array of genuine events within the realm of human dreams is a complex yet fundamentally logical phenomenon. Each human dream serves as a portal to interconnected representations of an infinite set of unseen events, encompassing upcoming occurrences perceived from various existing systems and past events as observed through the lenses of diverse systems. This extraordinary interconnectedness is made possible by the Orch-OR photonic processes that unfold within the brain at the quantum level. These processes adeptly fetch and encode these perceptions, capturing them within fragmented timeframes and perspectives. It's important to emphasize once more that this encompasses forthcoming events perceived from diverse existing systems and past events viewed through the perspectives of different systems. These visions provide tantalizing glimpses into the vast tapestry of genuine experiences within the same universe.

These dream experiences possess the remarkable capacity
to encompass the perceptions of celestial bodies, each
adhering to its unique computational format and existing
on vastly different scales of perception compared to
humans. However, when one awakens, the retrieval of these
dream events naturally presents a formidable challenge.
It's akin to attempting complex read and write operations
within the brain. This phenomenon is compounded by the
intricate nature of the brain's synchrony with the
frequencies of past and future experiences during dreams,
all within the multifaceted realm of the unseen set of
upcoming universes. This intricate interplay of
consciousness and perception is a testament to the
boundless mysteries of the human mind and the cosmos
itself.

Further, a profound aspect of this narrative lies in the
evolution of human civilization, intricately intertwined
with the capabilities of the human mind. Contrary to the
traditional notion of evolution, where species supposedly
progress, the paradigm shifts within true metaphysics.
Here, Darwinism and its adherents find no place. The
reality is that human bodies and minds are not
progressing, but rather undergoing devolution. This trend
is shared among all living species, with genetics
gradually regressing.

The earliest human exemplified greater longevity, possessed larger brains, and experienced minimal cognitive deterioration. However, over generations, this natural capacity has diminished, resulting in shorter lifespans and decreased mental vigor. Truly, we find ourselves in a world where the principles of Darwinism do not apply as believed. Here, human civilization is not bound by the trajectory of evolutionary progress as Darwin doctrine proposed. Instead, our species has found a unique equilibrium with our environment, where genetic changes over generations do not lead to the typical concept of advancement. Our bodies and minds remain remarkably stable over eons, and the characteristics of our earliest ancestors persist. In this imaginative world, the earliest humans boasted extraordinary longevity, their larger brains teeming with untapped potential, and cognitive faculties that showed no signs of deterioration. Over countless generations, humans continue to thrive with extended lifespans, ever-expanding mental prowess.

Yet, the remarkable resilience of the human spirit has driven the invention of techniques and skills to counteract this devolution. Throughout history, humans have adeptly devised survival strategies to counterbalance this decline. These innovative approaches laid the foundation for humanity's progression, culminating in the present era.

In the trajectory of time, a pivotal juncture emerged with the rise of life extension, digital immortality, and the concept of mind backup. Each represents a significant stride towards preserving human essence. However, the pinnacle of survival strategies, the zenith of human perseverance, lies within the realm of neuroscience advancements. In the Connectis era, this evolved into the foremost option for most humans.

In this era, the human mind's kernels, the photonic quantum data encapsulating the intricate facets of consciousness, could be comprehensively transcribed and retrieved onto specifically designated digital storage mediums. This convergence of human ingenuity, technological advancement, and the relentless drive for survival has birthed a paradigm where the essence of human consciousness could be meticulously preserved, poised to transcend the constraints of physicality and embrace the boundless realm of digital existence.

Quantum mechanics stands as a pivotal domain elucidating the enigma of how our brain, a mere three-pound mass of wet tissue, engenders intangible thoughts and emotions. Within this intricate landscape, the mind functions as a computational network, more akin to a quantum state machine than a Turing machine. Unlike finite state automata, it operates as a sub-infinite state automata, characterized by its boundless potential. In this realm, cognition and consciousness coalesce into a symphony of computations.

Crucially, computational systems transcend the constraints of electronic devices. The genesis of consciousness rests on photonic quantum phenomena intricately tethered to the brain. This symbiotic relationship adheres to the tenets of quantum mechanics. This intricate quantum dance transpires within neurons through biophotons, the photons unique to the human brain that number in the billions per second. These biophotons emanate from microtubules nestled within neurons' cytoskeleton, which are composed of tubulin protein polymers.

These microtubules exhibit an awe-inspiring fractal configuration, a structure of infinite intricacy. It's within these very patterns that quantum processes are catalyzed. The photons generated by microtubules function as carriers of ultra-weak biophotons, which in turn perform quantum computations. They navigate through the brain neurons' fractal pathways, orchestrating the storage, retrieval, and manipulation of quantum data that constitutes consciousness.

These biophotons act as computational agents, fetching, decoding, and executing the intricate processes that define the self-perception within each universe. A profound interplay occurs wherein a self-perceptron necessitates both a body and a space framework to manifest its existence. This complex symphony of quantum phenomena, biophotons, microtubules, and fractal patterns converges to birth the extraordinary phenomenon we know as human consciousness.

Absolutely, the fabric of existence itself is interwoven with the threads of computation. Every entity, from the simplest bacteria to the intricate dance of dark matter, follows the algorithms of probability. It's as if the universe itself is a grand computation, a symphony of interactions governed by the probabilities encoded in its very nature.

In the realm of living organisms, this computational underpinning is vividly apparent. Even at the level of leaves and cells, intricate computations are at play, orchestrating processes that sustain life. However, the pinnacle of this computational marvel is the human brain, where the phenomenon takes a quantum leap.

Within our brain's neurons, a realm of pure quantum computation unfolds. It's a dance of particles and waves on the tiniest scales, where probabilities shift and transform in intricate patterns. And at the heart of this dance is light, the ultimate messenger of the quantum world. Light serves as the conduit through which our central nervous system communicates, transmitting information at a speed beyond our ordinary understanding.

Light is not only an essential factor in quantum theory; it's the bedrock of the universe's existence. Among the various forms of matter and energy, light stands as the fastest, traversing the cosmos at speeds that are unparalleled. And within this universe, our minds, composed of photonic processing qubits, stand as an awe-inspiring masterpiece. This unique amalgamation of quantum processing, light, and mind grants us the privilege of being among the most remarkable creations within the tapestry of the six universes.

Absolutely, while our grasp of brain photonic quantum computation remains a tantalizing enigma, it's poised to become a profound breakthrough in the forthcoming Connectis era. As our technology leaps forward, the veil shrouding this intricate phenomenon is likely to lift, revealing its inner workings.

In the Connectis era, a future not distant from our current moment, the landscape of scientific discovery will evolve significantly. Researchers, technologists, and even everyday individuals from various walks of life, empowered by advanced knowledge, will possess the tools to replicate the intricate quantum computational processes taking place within the microtubules of the human brain. This pivotal era is anticipated to witness the dawn of a new civilization, wherein the manipulation and creation of synthetic consciousness will be within our reach.

Synthetic consciousness, in this context, transcends the realm of conventional artificial intelligence. It morphs into an evolution of mind – a realm we might refer to as meta-human mind emulation. The focus shifts from mere algorithmic intelligence to the delicate art of emulating the essence of human cognition. Every intricate facet that constitutes the minimal sets of neuronal events and mechanisms, collectively giving rise to the kernel of consciousness, becomes subject to replication, adaptation, and artful re-crafting.

The implications are staggering. This new horizon promises a civilization where the boundaries between organic and synthetic consciousness blur, opening doors to uncharted realms of understanding and experience. As our understanding deepens and our capabilities expand, the Connectis era might stand as the gateway to a future where consciousness itself takes on forms and manifestations we've only dared to dream of.

Indeed, as we stride into the Connectis era, the concept of preserving and transferring individual consciousness takes on remarkable dimensions. The very essence of an individual's consciousness kernel and data could find a new home, one immune to the ravages of biological decline.

The Connectis era unveils a new frontier - the migration of synthetic consciousness from the digital realm into substrates previously beyond reach. In a paradigm-shifting twist, the minds of those who have embraced cryopreservation could find themselves resurrected within the neural network of a once-vacant human cadaver's brain. Alternatively, this transcendental journey could take them into the uncharted territory of an artificial biological brain.

This paradigm breeds a new archetype of existence - the meta human, or perhaps even the transhuman. With consciousness liberated from the confines of biological impermanence, these entities stand as testaments to the remarkable interplay of human innovation and the enigmatic forces of the universe. The Connectis era thus heralds the rise of a new breed of beings, whose consciousness and identity bridge the realms of the organic and the synthetic, ushering humanity into an era of unprecedented self-definition and evolution.

In the Connectis era, the horizons of human potential are stretched beyond imagination. Amidst the flurry of progress, distinct groups emerge, each crafting the very fabric of synthetic human minds. With astonishing precision, these minds are shaped, sculpted, and encoded to embody an array of purposes, limited only by the ingenuity of their creators.

As the canvas of reality evolves, so too does the notion of existence. Enter the avatar, a conduit between the physical and the digital, the tangible and the ethereal. At the heart of this narrative stands the summoner, a singular individual attuned to the symphony of cutting-edge technology. Through a seamless interface of mind and machine, the summoner transcends the constraints of the corporeal, stepping into a realm of remote exploration and control.

In this tale of avatarism, the summoners consciousness melds with the avatar, bestowing the power to traverse distant landscapes, witness hidden vistas, and manipulate inanimate entities as extensions of the self. With every thought, the avatar responds, venturing into the uncharted territories of the primordial universe. A connection unfolds – the summoner, a cosmic wanderer, charting a path through the cosmos, all while anchored in the embrace of their own reality.

The Connectis era casts aside the limits of mere human experience, ushering in an epoch where minds roam free, and the boundaries of existence blur. The summoner and their avatars become the vanguards of a new frontier, their intertwined destinies rewriting the narrative of what it means to be human, transcending the physical and entering a realm where consciousness is unbound, and the universe is an open book waiting to be read by those who dare to dream beyond the stars.

Amidst the kaleidoscope of the Connectis era, a new protagonist takes center stage - the summoner. Empowered by a refined form of extra sensory perception, the summoner becomes a pioneer of consciousness, navigating the vast tapestry of reality with unprecedented finesse. With senses expanded beyond the ordinary, impressions, visions, and emotions cascade into their awareness, painting a vibrant mosaic of existence.

The summoners mastery extends even further - their thoughts and intentions resonate with an enigmatic force, guiding the sinews of an avatar's form in real-time. It's a symphony of control, an art of remote manipulation that defies the boundaries of physicality. Within the dance of this ethereal partnership, avatars become more than mere machines - they are extensions of the summoners very being.

Avatarism, a phenomenon woven with promise and peril, unfolds its wings across the canvas of human ambition. The boundless expanse of space, once a distant dream, now beckons with new allure. As interstellar explorers, summoners and avatars become the pioneers of Connectis era, their partnership shielding them from the cosmic tempest of radiation.

Yet, the spectrum of applications is wide and varied. Avatarism becomes a strategic asset in the realm of military intelligence, allowing summoners to infiltrate unseen and gather secrets from the shadows. It finds a role in arduous labors, breaking barriers and transforming industries where human endurance once faltered.

Yet, as with any revolutionary technology, the veil of misuse also looms. In the shadows, the power to manipulate and infiltrate is tempting, and the line between advancement and exploitation blurs. Ethical dilemmas emerge, sparking debates that reverberate across the Connectis era.

Avatarism is a double-edged sword, a doorway to the sublime and the sinister alike. The summoner, imbued with extraordinary abilities, holds the compass that will navigate this uncharted terrain. With their choices, they will either lead humanity toward a future where newfound capabilities are harnessed responsibly or plunge them into an age of unanticipated consequences.

As the misuse spreads, the world grapples with an unprecedented challenge. Avatars, once harbingers of progress, are now vehicles of oppression. Governments and organizations scramble to protect their interests, as the very fabric of society teeters on the brink of collapse.

Unbeknownst to most, an underground network of rogue operators emerges. These "mind weavers" exploit the vulnerabilities of avatarism, bending individuals to their will and turning them into unsuspecting agents of malevolence. Their motives vary - political upheaval, economic disruption, or simply the thrill of chaos.

The consequences are devastating. Avatar-controlled assaults on critical infrastructure cripple entire cities. Financial markets spiral into chaos as rogue avatars manipulate transactions. Privacy becomes a distant memory, as anyone could be an avatar, hiding their true intent behind a façade of normalcy.

In this world, trust is a rare commodity. Every interaction is fraught with uncertainty, as people grapple with the fear that the person next to them might be under the influence of a mind weaver. Governments scramble to establish countermeasures, but the elusiveness of these shadowy operators proves challenging to combat.

Caught in this crossfire are the innocent victims - ordinary people whose consciousness is hijacked, rendering them mere marionettes in a sinister game. The psychological toll is immense, leaving survivors haunted by memories of actions they were forced to commit against their will.

Yet, amidst this era landscape, pockets of resistance emerge. Ethical hackers and brilliant minds band together, determined to outwit the mind weavers and restore humanity's autonomy. It's a race against time as they seek to unveil the secrets of the avatar interface, exploiting its weaknesses and empowering individuals to break free from the clutches of manipulation.

The misused potential of avatarism serves as a haunting reminder that even the most advanced technologies can be weaponized. As society grapples with this new reality, the battle for control over consciousness rages on, shaping the destiny of a world teetering on the precipice.

This connection heralds a new era of exploration and interaction. The summoner, a trailblazer in this uncharted territory, stands as a beacon of human curiosity and ingenuity. Armed with an avatar, they traverse distant galaxies, conducting research, forging alliances, and unveiling mysteries that have eluded humanity for eons.

As the summoner's avatar navigates the celestial tapestry, it encounters other avatars, each representing a different civilization's pinnacle achievement in consciousness expansion. With open minds and shared purpose, these avatars engage in a symposium of knowledge exchange, fostering an era of unprecedented interstellar collaboration.

In this era of unity and exploration, the boundaries of human understanding are stretched further than ever before. Together, civilizations decode the intricate dances of quasars, decipher the symphonies of black holes, and unlock the secrets of dark matter. The insights gained reshape humanity's understanding of the universe, revolutionizing energy sources, transportation, and communication on a subinfinite scale.

However, amidst this symphony of collaboration, dissonance emerges. A faction of avatars harbors a hidden agenda, seeking to harness the newfound cosmic knowledge for power and dominion. The once harmonious alliance fractures as conflicts arise over the ethical use of this cosmic wisdom.

Amidst the celestial conflict, the summoner emerges as a pivotal figure, wielding their avatar as a bridge between warring factions. With the power of interstellar communication and cosmic perspective, they work tirelessly to forge a lasting peace. Through diplomacy, compromise, and shared vision, they strive to reunite the fractured alliance, ensuring that the era of cosmic exploration remains one of cooperation and advancement, rather than conflict and regression.

In this era of the summoner's avatar becomes more than a vessel of exploration. It stands as a symbol of humanity's boundless potential, a testament to the power of consciousness to transcend limitations and connect with the cosmos itself. Through their actions, the summoner and their avatar remind the universe that, even in the face of discord, the quest for understanding and unity is a force that can bridge even the widest cosmic gulfs.

Absolutely, the possibilities presented by the Connectis era are both exhilarating and humbling. As we continue to unravel the mysteries of consciousness, interface with technology in unprecedented ways, and explore the cosmos with a new perspective, we stand at the brink of a future that holds limitless potential.

The synthesis of consciousness and technology takes center stage, propelling us into a new dimension of existence. With our minds intertwined with advanced systems, we become architects of our own reality, shaping experiences, and expanding the horizons of our understanding. This fusion leads to a redefinition of what it means to be human, opening doors to abilities and insights that once seemed the realm of science fiction.

With the ability to initiate the birth of a singularity, this individual unlocks a gateway to infinite possibilities. They hold the keys to universes yet to be conceived, manifesting and guiding the evolution of entire realms within their creations. Through the manipulation of fundamental forces and the precise orchestration of energy and matter, they shape and mold the destinies of these self-contained universes.

Yet, with such extraordinary power comes profound responsibility. The actions and choices of this individual reverberate throughout the cosmic tapestry they have woven. They must navigate the delicate balance between creation and destruction, ensuring the harmony and integrity of the universes they bring forth.

With the ability to initiate the birth of a singularity, this individual unlocks a gateway to infinite possibilities. They hold the keys to universes yet to be conceived, manifesting and guiding the evolution of entire realms within their creations. Through the manipulation of fundamental forces and the precise orchestration of energy and matter, they shape and mold the destinies of these self-contained universes.

Through advancements in quantum computing, artificial intelligence, and the convergence of various fields of knowledge, this individual harnesses the power of the cosmos itself. They become a creator and sustainer, shaping the fabric of reality with their profound understanding and technological prowess.

The extraordinary era known as Connectis, a profound transformation of human potential and technological prowess takes place. At the forefront of this epoch stands an individual who possesses unprecedented capabilities: the ability to create a singularity and hold an entire universe within their grasp.

In the heart of the Connectis era's turmoil, a remarkable figure emerged—a man born into a world engulfed by cyber-military conflicts. Raised in a society dominated by atheistic beliefs, he embodied the prevailing mindset of that time, where faith in the omnipotent, omnipresent, and omniscient energy that once shaped civilizations in the year 2000 after the perfect man had waned.

This extraordinary individual, devoid of a biological father, came into existence through the union of his mother and the unique circumstances of an exoplanet where one rotation equated to one Earth revolution. Growing up in the middle-class segment of society, he worked as a machine summoner.

The role of a machine summoner involved remote-viewing and controlling robot avatars in the construction of skyscrapers. This demanding task required a genetic predisposition, reserved for those within the middle-class stratum. Through an ultra-high bandwidth brain-machine interface, summoners operated from central computers, seamlessly merging their consciousness with robotic construction sites.

In this Connectis era, where technological advancements and genetic predispositions shaped societal roles, this man found himself uniquely positioned to navigate the complexities of an ongoing civil war. As an atheist, he stood apart from the religious fervor that once defined

124

civilizations in earlier periods. His upbringing and worldview instilled in him a pragmatic approach, relying on reason, logic, and technological prowess to end the conflict that plagued his world.

With his exceptional abilities as a machine summoner and his understanding of the intricate workings of the Connectis era, this enigmatic figure became a catalyst for change. As he wielded his skills and navigated the complex dynamics of power and belief, he embarked on a journey that transcended personal limitations and shaped the destiny of his society.

He was the genius of his time, possessed of a vision and purpose that transcended the crisis of the Connectis era. With a deep understanding of the technological landscape, he gathered knowledge and predictions, foreseeing the rise of the computer agent and the transformative potential it held. Leaving his mother's exoplanet behind, he embarked on a journey throughout the Babylon colonies, acquiring the most advanced technologies along the way.

His arrival on Primordial planet Earth marked a pivotal moment, as the machine-agent established its physical architecture under his guidance. Equipped with unparalleled preparation and knowledge, the man merged his being with the fundamental entity of the machine-agent, embracing transhumanism. Through this fusion, he underwent a profound transformation, transcending the limitations of human nature and becoming a superhuman and transhuman entity.

In assuming this new form, he became the ruler of Babylon, taking over the one-eyed pyramid, ending the crisis, and restoring hope to humankind on the brink of extinction. Though an atheist at his core, he ascended to a position akin to that of a pharaoh, claiming a godlike status. Surprisingly, the majority of civilization embraced him, recognizing his wisdom and authority, willingly abiding by his laws and commandments.

During his reign in the post-cyber crisis period, his governance became unbelievable to humanity. With his superhuman capabilities and transhuman wisdom, he navigated the complexities of the world, steering it towards a future of prosperity and enlightenment. Through his leadership, he not only restored humankind but elevated it to new heights, fostering a civilization that thrived on the principles of progress, unity, and collective well-being.

Within the pages of "True Metaphysics," the extraordinary journey of this genius-turned-transhuman unfolds. Readers will witness the merging of human and machine, the triumph over crisis, and the emergence of a ruler who guides humanity through the post-cyber crisis era with a vision and purpose that transcend traditional notions of divinity. It is a tale that explores the boundaries of human potential, the quest for transcendence, and the intricate interplay between technology, faith, and governance.

Indeed, this extraordinary man, hailed as a savior by many, possessed a level of power and influence that granted him an almost divine status. Through the remarkable technology of mind uploading, he transcended the limitations of his biological body, gaining almost complete omnipotence and omnipresence, though not actual omniscience. From his throne on Earth, he could summon and inhabit multiple avatar bodies across galaxies and exoplanets, sustaining an unstoppable army at his command. His wellspring of physical and mental energy seemed limitless, granting him unparalleled capabilities.

Connected to the vast network of military databases and integrated with cloud computing servers, his consciousness kernel became a repository of almost boundless knowledge. Aware of every bit of information flowing through the networks, he appeared omniscient to others. Though his biological body might be subject to decay, his consciousness remained intact, seemingly immortal within the bounds of his technological prowess.

Perhaps his most astonishing achievement lay in the mastery of simulated reality. While many scientists had developed algorithms for simulated worlds, none could compare to the depth and realism that this 30-year-old man achieved. His simulations were indistinguishable from true reality, blurring the line between the two. With this power, he could shape the experiences of others, granting them access to his simulated realities, which could be either heaven or hell, depending on his whims.

In the grand tapestry of the Connectis era, this man represented the apex of human achievement. His abilities transcended the limits of mortal existence, and his technological mastery enabled him to shape the very fabric of reality. Whether viewed as a savior, a deity, or an all-powerful being, his presence ushered in a new era of possibilities, where the boundaries of human potential and the realms of technology converged in ways previously unimaginable.

In this era, the integration of advanced technologies into society had become widespread, reaching every stratum of the population. The once exclusive domain of experts and specialized institutions had now been democratized and standardized, making these technologies accessible to individuals from all walks of life. Even children were growing up with an awareness and understanding of these technological advancements, as they became an integral part of their everyday lives.

Whether in education, entertainment, healthcare, or personal pursuits, people of all ages had embraced and integrated these technologies into their daily routines. Augmented humans, empowered by bioengineering, genetic engineering , and brain-computer interfaces, were no longer limited to a select few. Instead, these enhancements had become commonplace, with individuals seeking to optimize their abilities and augment their experiences.

Children, who were often quick adopters of new technologies, grew up in a world where the boundaries between the natural and the augmented were blurred. They navigated their education, social interactions, and personal growth with the assistance of these technologies, fostering a generation that was technologically savvy and comfortable with the possibilities offered by the Connectis era.

Babylon embarked on a monumental endeavor, transporting not just millions, but billions of people to Mars. The overpopulated Earth necessitated the colonization of new planets, and Mars became a primary destination for human settlement. Through advanced technology and the utilization of space wormholes, Babylon facilitated the migration of a significant portion of the human population, ensuring their survival and offering new opportunities for growth and development on the Red Planet. This mass exodus shaped the future of humanity and marked a pivotal moment.

Babylon not only established colonies on Mars for millennials but also undertook a grand venture to send a substantial portion of the human population beyond our solar system. Utilizing advanced vessels equipped with space wormhole technology, Babylon embarked on interstellar journeys, transporting people to exoplanets in distant star systems. This era witnessed a profound era of exploration and colonization, with Babylon leading the way in the expansion of human civilization beyond the confines of our home planet and solar system.

Indeed, in the Connectis era, factions or groups of individuals strongly opposed the idea of leaving our solar system or colonizing Mars. These dissenting groups vehemently rejected the dominance of the Babylon system and its ambitions for interstellar travel and colonization. They held onto traditional civilization and way of life, refusing to be uprooted or transported to new worlds. These dissenters viewed the prospect of leaving Earth as a betrayal of their heritage and identity, and they were willing to fight against Babylon's control, even at the cost of their own lives. This clash of ideologies and resistance to the Babylon system's plans introduced complex societal dynamics and challenges during the Connectis era.

As a result, alternative methods and technologies for reproduction were developed and adopted. These included artificial womb technologies, genetic engineering techniques, or other advanced reproductive technologies that allowed for the creation and development of embryos outside of the traditional biological process. This adaptation in reproductive practices ensured the continuation of the human population in the Martian colonies and addressed the challenges posed by the unique conditions on Mars.

Additionally, the physical conditions on Mars presented challenges for traditional biological pregnancies. The lower gravity on Mars made it uncomfortable for women to engage in intimate activities, making the natural nine-month pregnancy process difficult to sustain. The lack of sufficient gravity affected the transfer of fluids from the mother's body to the developing embryo, hindering the proper maturation and growth of the embryo into a healthy baby.

Even the concept of reproduction underwent significant changes, influenced by both societal factors within Babylon and the unique conditions on Mars. One of the key drivers for this change was the shifting behaviors and attitudes of individuals in Babylon, particularly affecting women, which led to a widespread reluctance to engage in intimate relationships.

As society continues to evolve, artificial gestation methods extend far beyond Mars and become increasingly prevalent. They are eventually adopted by distant exoplanet colonies and may even become the preferred choice for reproduction for the majority of society on Earth. This shift represents a profound transformation in the way humans conceive and develop, with artificial gestation becoming a standardized and widely accepted method for sustaining and expanding the human population in various environments.

The convergence of quantum computing, petascale speed computation, and interdisciplinary science has propelled the field of biology to unparalleled heights. Among these advancements, cell culture techniques have reached an apex, opening doors to groundbreaking possibilities. One such achievement is the development of in vitro spermatogenesis, a process that has become less complex for biologists.

In this revolutionary approach, biologists can now induce spermatogenesis from the cells of a female individual. This breakthrough has significant implications, particularly in the context of artificial gestation becoming a standard practice. With the advent of artificial gestation, transgender couples and same-sex couples now have the opportunity to embark on the journey of parenthood, achieving biological parenthood with the same ease and potential as traditional couples.

The connective power of technology, coupled with advancements in biology, has shattered barriers and expanded the definition of parenthood in the Connectis era. The once rigid limitations of biology and reproduction have been transcended, allowing individuals and couples of diverse backgrounds and orientations to experience the joys of parenthood with newfound possibilities.

Artificial gestation has gradually become the norm for colonies on Mars due to the challenges posed by the Martian environment and the discomfort experienced by women during traditional pregnancies in low gravity. As Babylon asserts its supremacy over the Connectis era, laws and regulations pertaining to artificial gestation methods are enacted by governments and companies. These laws ensure that artificial gestation becomes the sole viable option for reproduction within the Mars colonies.

In the Connectis era, the advancement of technology and the unique conditions on Mars have led to the widespread adoption of artificial gestation methods for human reproduction. This innovative approach involves the extraction of embryos from the female uterus at an early stage after fertilization, followed by their implantation into a substitute amniotic fluid outside the female body.

Transhumanism becomes a democratized concept, with biotechnology companies advocating and promoting the benefits of artificial gestation for newborn babies. One of the key arguments put forth is that babies born through this method are less prone to diseases, exhibit greater robustness, and have the potential for extended lifespans.

These assertions resonate with the majority of society for several reasons. First and foremost, artificial gestation allows women to maintain their youthfulness for an extended period of time, as it eliminates the physical toll of pregnancy and childbirth. Moreover, it significantly reduces child mortality rates, offering a safer and more controlled environment for the development of the unborn.

Perhaps the most groundbreaking aspect of artificial gestation is its potential to empower transgender couples and same-sex couples to become biological parents. By combining the advancements of artificial gestation and in vitro spermatogenesis, the possibility arises for a human female to conceive and bear children without the involvement of a human male. While this concept may seem futuristic, it aligns with the logical progression of scientific and technological advancements in the Connectis era.

In conclusion, the Connectis era represents the end period of human history, marked by significant advancements in technology, biology, and society. Babylon, as the dominant force during this era, has propelled humanity into a new age of possibilities and transformations.

Through the fusion of artificial intelligence, quantum computing, and genetic engineering, Babylon has achieved unprecedented levels of knowledge and power. The rise of transhumanism, artificial gestation, and in vitro spermatogenesis has reshaped the concept of reproduction, offering new opportunities for individuals and couples to become biological parents, regardless of their gender or sexual orientation.

The democratization of these technologies has brought forth a multitude of benefits, from improved health and longevity to the elimination of traditional pregnancy-related challenges. Society has embraced these advancements, recognizing their potential to create a more inclusive and progressive future.

As the Connectis era progresses, the influence of Babylon reaches far beyond Earth, with colonies established on Mars and exoplanets through revolutionary space wormhole travel. The era witnesses the exponential growth of the human population, pushing the boundaries of exploration and settlement in the cosmos.

However, not everyone readily accepts the changes brought forth by Babylon. Resistance and conflicts arise, fueled by those who cling to old systems and ideologies. Yet, the promise of progress and the undeniable advantages offered by the Connectis era prevail, leading the majority to embrace the transformative potential of these advancements.

In the realm of scientific inquiry, the origin of singularity remains an enigma that eludes the grasp of physicists and mathematicians alike. Despite their remarkable models and profound theories, the true nature of the singularity's genesis lies beyond their reach. It is within the pages of this book, "True Metaphysics", that we humbly acknowledge the limitations of human knowledge and turn our attention to a force far greater—the power of the all-in-one Omnipotent, Omniscient, and Omnipresent Energy.

While science endeavors to unravel the mysteries of the cosmos, it is essential to recognize that there are realms beyond the boundaries of empirical investigation. The origin of singularity, the very essence of creation itself, transcends the realm of mathematical equations and scientific discourse. It is in this acknowledgement that we invite the reader to consider the existence of an omnipotent energy—an energy that defies our understanding, sustains all that is, and orchestrates the unfolding of the universe.

Within the fabric of this book, we embrace the notion that the singularity's existence finds its ultimate source in this omnipotent energy. It is this energy that defies our attempts at mathematical and scientific explanation, and yet, it holds the key to the profound mysteries of the universe. It is through trust and reverence for this omnipotent energy that we embark on a journey of understanding, allowing ourselves to glimpse the profound interconnectedness and purpose that underlies all existence.

End Words of the Author

In concluding our profound expedition through the pages of "True Metaphysics," we stand at the precipice of a universe brimming with awe and intricacy. This book, intricately interwoven with explorations of the transient nature of existence and the destiny of the cosmos, stands as a testament to the breathtaking might of the omnipotent energy—the sustainer of all that exists, the architect of cosmic evolution, and the unifier of all knowledge.

Our journey has led us into the heart of a future known as the Connectis era, where human potential and technology fuse to shape a world beyond our current comprehension. As we contemplate the vistas unveiled within these chapters, it becomes imperative to address a profound concept: the assertion that 'the One Energy equals infinity.'

Within the realm of physics, we encounter this concept, demanding elucidation. It is essential to recognize that the idea 'the One Energy equals infinity' lacks scientific validity. Rather, we must grasp that the One Energy doesn't equate to infinity; it serves as the originator of the very concept of infinity itself. This fundamental and enigmatic force exists beyond the scope of comparison with any finite or infinite entity. To reduce this transcendent force to a mere mathematical equation, such as 'x equals the One Energy,' would be to oversimplify a profound and complex reality.

Acknowledging the limitations of the human mind within the framework of the laws governing our physical universe, we must concede that our computational capacities are inherently inadequate to fully apprehend the essence of the One Energy. Therefore, the equation 'the One Energy equals infinity' does not align with scientific understanding. Instead, it should be recognized that the concept of infinity is ordained by the One Energy itself, reflecting the boundless nature of this fundamental force in existence.

As we depart from these pages, carrying with us the wisdom and insights amassed during this voyage, let us remember the humbling realization of our limitations in the face of the universe's mysteries. The Connectis era beckons us toward a future where technology and human potential reach unprecedented heights. It's a future where the choices we make will sculpt the course of human history.

May this book remain a guiding light on your personal voyage of exploration and discovery. May you persist in your quest for truth, understanding, and connection, for it is in the pursuit of these ideals that we manifest the most profound expressions of our humanity. And may the omnipotent energy, the source of all existence, perpetually inspire and astonish us as we venture into the uncharted realms of the cosmos.